High Mysticism

High Mysticism

EMMA CURTIS HOPKINS

COSIMOCLASSICS

NEW YORK

High Mysticism
Cover © 2007 Cosimo, Inc.

For information, address:

Cosimo, P.O. Box 416
Old Chelsea Station
New York, NY 10113-0416

or visit our website at:
www.cosimobooks.com

High Mysticism was originally published in 1888.

Cover design by www.kerndesign.net

ISBN: 978-1-60206-210-8

Mistakes of mind and action may be conscious or unconscious
in the part of mankind. When mistakes are unconscious people
may never trace the mechanical consequences of their mistakes in
the misfortunes of daily life. The mother compels the child to study
his lessons, not knowing that his eyes or brain may be weak,
and in after years, when he is insane or blind, she is totally unconscious
that she had once pressed his brain or eyes beyond their bearing point.

—from *High Mysticism*

CONTENTS

I

REPENTANCE—THE SILENT EDICT

From the Divine Heights there has been vouchsafed to all ages One Heavenly Edict. All the everlasting pages struck off by men under the white flames of inspiration, have been the result of knowing or unknowing obedience to the Soundless Mandate of the Lofty One inhabiting Eternity:

"Look unto Me, and be ye saved, all the ends of the earth."

A clearly unified instruction runs in almost verbatim language through all the sacred or charmed books of the world. It is the live wire insulated by absurd dogmas and ungodly imaginations. It is the footpath of the immortals. It is the mirific science. Whoever can read its supernal lines, undiverted by their company of errors, is in the way of salvation. It is that swift, subtle faculty possessed by us all, whereby we look whithersoever we will, to the Deity ever beholding us, or to the dust beneath, without the aid of our physical eyes.

"Thou canst not behold Me with thy two outer eyes, I have given thee an eye divine."—*Upanishads.*

This fleet, subtle sense is our incorporeal eye. It is the one faculty of our immortal soul which we continually make use of. It is the creature made subject to maya, not willingly, but in the hope of redemption of the body, as Paul wrote to the Roman Christians.

The exaltation or lifting up of this sense toward that Vast, Vast Countenance ever shining toward us as the

1

sun in his strength, is our way of return to the Source whence we sprang forth. It is the Path of Light. It is the Tao.

> "Make use of the light, returning again to its Source;
> Thy body shall be free from calamity's course,
> And thou shalt train with the Eternal at length."
> —*Tao-teh-King.*

"Man alone of all the animals goes in quest of his Origin, and perceiving that the highest good is to be sought by him in the highest place, looks to his Maker."— *Lactantius.*

This looking faculty antedates mind, and though offering itself to the service of mind, transcends it in achieving power. For it is primarily what we most see, and not what we most think, that constitutes our presence, power and history.

"It is not possible for anything to take place save in connection with an onlooker," reads an inspired line in the Vedic Hymn.

If we exalt this swift sense, or look unto Him whose ever repeated mandate is, "Behold Me, behold Me," we receive back over the track of our vision tonic and viability to the mind, endurance and beauty to the body, joy and fearlessness to the emotions, integrity and intrepidity to the moral character.

All that we think is made up of the objectives toward which we have directed this deathless, achieving visional power. All that this posit we call body exhibits is the set of accretions that has come over the inner visional track.

"That thou seest, that thou beest."

We collect sadness and depression from directing this

mystic eye toward human faces. Because of this manner of attention did Solomon weep so loudly his retainers trembled. Sanity and soundness are the characteristics of the mind of those who do not project their prehensile vision toward objects that gratify the five outer senses. They who see toward the Heights are invulnerable to honor or contempt, praise or dispraise. Their probity, sincerity and courage lapse not.

> "For that thou seest, man,
> That too become thou must;
> God if thou seest God,
> Dust if thou seest dust."

With closed eyes, still let the gaze be heavenward; there on the fair unspeakable Heights is the home whence we all came hitherward to view the ways of destruction:

"Thou turnest man to (see) destruction; and sayest, Return, ye children of men."—*Psalm of Moses.*

To look upward with the mystic eye is to start on the saving Tao. "Look unto Me, and be ye saved—I will turn away your captivity from before your eyes—when ye turn unto Me seeking My face," declared the two great prophets, Isaiah and Jeremiah.

"With the flash of one hurried glance I attained to the vision of that which Is. And Thou didst not give me any peace till Thou wast manifest to the eye of my soul," cried St. Augustine of Tagaste, in one of his illuminated moments.

The farther toward the celestial zenith we send the limitless eye, the deeper is our assurance of our own divine origin and transcendent Selfhood. For truly the Highest is the Nearest, the most distant yet most present, and we are in His image.

"The Highest and the Inmost are one," declared the two great mystics, Behmen, and Hugo of St. Victor.

> "Look up, my comrade!
> When on the glances of the upturned eye
> The plumed thoughts take travel, and ascend
> Through the unfathomable purple mansions,
> Treading the golden fires, and ever climbing,
> As if t'were homewards winging—at such time
> The native soul, distrammelled of dim earth,
> Doth know herself immortal, and sits light
> Upon each temporal place." — *Violenzia.*

"If then there be any incorporeal eye, let it come forth from the body, and to the Vision of the Beautiful. Let it fly up and be lifted into air; not figure, not body, not ideas, seeking to contemplate, but that rather the Maker of these: The Quiet, The Serene, The Stable, The Invariable, The Self, all things and only The One; The Like to Itself, which neither is like to another."—*Cyril, Bishop of Alexandria.*

In high moments of recognition of the light that transcends reason, man transcends himself, and writes more wisely than he knows.

"No man when in his wits attains prophetic truth and inspiration, but when he receives the inspired word his intelligence is enthralled."—*Plato.*

Lifting the inner eye to Him who is above reason lights the two outer eyes to see the world in a new aspect, gives the tongue new descriptions of the world, and tips the pen with fadeless phrases. And that descending light, compelling transformation of all surrounding objects, is the mystic river of which the angel told Ezekiel, ". . . everything shall live whithersoever the river cometh, and everything on its banks shall be healed."

The healing of the mind to think supernal truth waits upon that light which only the uplifted mystic eye can bring to mind. The transfiguration of matter waits upon the flawless ecstasy which only the mystic eye can find. Order and beauty hide their sublime mysteries till on the Tao's magic path the tireless vision speeds toward the Origin of beauty and order.

"In heaven there is laid up a pattern which he who chooses may behold, and beholding, set his own house in order. The time has now arrived at which they must raise the eye of the soul to the Universal Light which lightens all things. With the eye ever directed toward things fixed and immutable which neither injure nor are injured—these they cannot help imitating. But I quite admit the difficulty of believing that in every man there is an eye of the soul which by the right direction is re-illumined, and is more precious far than ten thousand bodily eyes."—*Plato.*

As down the sides of Hermon, the unapproachable, trickle cooling dews to refresh the hot valleys, so falls a reviving miracle of newness upon the children of earth when they penetrate beyond the stars to Him who proclaimeth forever, "Behold, I make all things new."

As balm from the trees of old Gilead in far past days soothed the hurts of the Jews, so the Dayspring from on high doth visit them that sit in darkness and in the shadow of death, to guide their feet into the way of peace. Nothing we can do, or say, or think can quench the downfalling reconciliation and empowerment, the preserving and healing, while to the High Edict responsive we lift up our eye to the hitherward smiling Countenance of the Lover ever with us, the Lord of Hosts His name.

He, abiding as the Great Different gives peace which

nothing can invade. His benedictions confer resistless might. Therefore, "Behold, as the eyes of servants look unto the hand of their masters, so our eyes wait upon the Lord our God. . . ."

This deathless visional faculty is our only achieving power. It is not dependent upon thoughts of mind or bodily actions, though to them it yields itself day by day in omnipotent servitude. Left to itself it flies away to the Elysian Fields, its rightful resting place.

So eagerly did the untaught seers of the past long to have this immortal faculty find its rightful direction, they willingly practiced mortifications of the body, denied self, affections as well as appetites, to give it freedom. But it asks no such sufferings on the part of the mind or body to give it power to tame and glorify them. It asks only their will that it go homeward.

It is the immaculate of us. Though age and decrepitude have cramped the flesh, senility has sapped the mind, and sickness has blinded the eyes and thickened the ears, yet the wrecked old man lifts up his sightless eyes and smiles. With the immortal and ever young mystical eye he beholds things celestial. And then he drops the robe of clay, hastening to be identified with his joy-giving vision. Had the eye been lifted to the mountains of help in earlier days, he would have transfigured and renewed his flesh, instead of leaving it to the moth and the sod.

All the other faculties in daily use are maculate. The mind can become vitiated, the body can become diseased, but though this all-accomplishing sense can bring back on its beams the nature of that upon which it may be stayed, itself has received no tinge of similitude . . . the same out of itself, the same in itself—*a-se-ity*.

With this all-accomplishing sense we are to repent—

to return. ". . . but now (God) commandeth all men everywhere to repent," declared Paul to the Athenians.

"Repent . . . and turn away your faces from all your abominations" was Ezekiel's admonition.

"The eye of the soul, which is literally buried in an outlandish slough, is by right science lifted upwards."— *Plato.*

And this is that return which hath divine reward: "Return unto me, and I will return unto you," said the heavenly voice to Malachi.

The mind cannot return, "For as the heavens are higher than the earth, so are my thoughts higher than your thoughts." The footsteps of flesh cannot return, "For as the heavens are higher than the earth, so are my ways higher than your ways."

"For these are but the distance of the strengthless from the Stronger, the short-lived from the Eternal, and the phantasy from the Like-to-Itself-Only."—*Hermes Trismegistus.*

But the heavenly vision rests her fleet splendor in the high Source from which the flawless and immortal soul sprang forth:

"I have given thee an eye divine with which to behold My power."—*Upanishads.*

By turning the celestial faculty toward the heights we are taken above the thought circuit to the watch:

"Watch ye therefore"—"What I say unto you I say unto all, WATCH."—"Blessed are those servants whom the Lord when He cometh shall find watching."

All miracle workers have practiced the principle of watching. Moses, the genius in leadership, speaks unto the nation of slaves:

"Stand still, and see the salvation of the Lord, which

he will shew to you." For the "Lord shall fight for you,
and ye shall hold your peace."

And this is forever, inevitably, the prayer of the
supernally inspired leader of men: "Look down from
Thy holy habitation, from heaven, and bless Thy people. . . ." And they shall pass in safety through Red
Seas of difficulty, though all the powers of mind and
matter oppose them.

And this is forever the joyous chant of the liberated
people: "He looked on our affliction, and our labor, and
our oppression, and He brought us forth out of darkness,
with a mighty hand and with an outstretched arm."

Elisha the seer stood with gaze transfixed toward a
seraphic host in the mountains round about Samaria.
"They that be with us are more than they that be with
them," he said to his servant who watched with him.
And though the king of Syria had sent soldiers to slay
the lonely prophet, they were not able to hurt him, for
mystic defense transcends the sharpest swords.

Is it not promised: "I will give power unto My two
watchers"—new powers, miraculous powers!

St. Bernard, abbot of Clairvaux, rose to almost supreme power in his church, by persistently gazing toward
the twelve stars in the diadem of Mary in Paradise. He
urged others to do likewise: "If the winds of temptation
blow fiercely upon you, look to these stars. If you find
yourselves in a sea of trouble, look to these stars. If you
are tossed on the waves of pride, ambition, envy, look
to these stars, and invoke the name of Mary."

Earlier, Hosea, making note of such as St. Bernard,
cried, "They return, but not to the Most High."

Savonarola pictured before his inner eye a monastery
for a holy resting place from turmoil and strife. Its

monks should all be men come not to be ministered unto, but to minister. And it is recorded that so influential did the outcome of his vision grow, that great citizens begged to join the Dominicans, and riotous processions, idle songs and fightings ceased on the streets of beautiful Firenze. With his inner eye on the commanding form of a warning visitant from the shores of mystery, Savonarola drew order out of chaos, and established a new form of government in the city of the Medici. In a time of dearth and danger, loaded wheat ships arrived, and the enemy's troops were not able to reach the people under the protecting ministry of Savonarola the seer.

All the forces of the universe cooperate with vision toward beatific ideals. It is not till the eye descends to prowl among the viciousness and crimes of men that war and martyrdom succeed. So descending did Chrysostom, the golden-mouthed, forget to show the glories of the heavenly land, and he perished in exile. Jeremiah lamented so profoundly over the mistakes of the Jews that he was martyred in Egypt. Elisha never lost his high watch, and even his bones were life-giving. His whole pathway on earth was strewn with miracles. For no weapon formed against the comrade of angels can prosper—radiating forever what he assimilates.

Hufeland secretly eyed the unspoilable region of spiritual health in his diseased patients, and they recovered. The Hidden Actual readjusted the molecules and atoms of the manifest, to harmonize with Hufeland's untaught visional practice. Gordon noted that those who reported to him their procedure while demonstrating miraculous cures, mentioned seeing with their inner eye some gesture or image symbolic of, or identical with, the healing about to show forth. Maxwell watched the fleet, ethereal light which he discovered filling all quarters of the universe, and he declared that by watchful use of it the ailing among mankind might all be made whole.

The difference between the great men whose names have attracted the attention of mankind, as to endurance in memory, and strength in perpetuating their doctrines, has depended upon the uplift they have given to the hidden eye whereby the mind receives elation or depression.

Socrates came not to teach any positive doctrine, but to convince man of the ignorance of his mind. His highest science got no higher than that men act wrongly because they form erroneous judgments. Upon being told that he was the wisest man, he said it probably was true, for he knew enough to know that he knew nothing, while no one else seemed to know that much. The ignorance of a man's mind is a dark zone to fix the all-collecting eye upon. No joyous inspiration fulgurates from that Ethiopic field.

Gautama Buddha cried, "I will now seek out a noble law, unlike the worldly methods known to men. I will oppose the scourges of the world, old age and death, disease and poverty." And at last he proclaimed that in order to be blest, man must keep eight conditions, and the first is right view: "For it is not possible for anything to take place save in connection with an onlooker." Thirty thousand miracles of achievement followed in his wake, and one-third of the human race hold him and his sayings in loyal reverence to this day.

The world-conquering Jesus crowned the doctrine of the exaltation of the supernal sense with immediate demonstrations:

"Father, I will that they may behold my glory." And

multitudes came unto Him, and He healed them every one. To the blind man with the clay upon his eyes He said, "Look up." To all people in times of calamity, He said, "Look up, for your redemption draweth nigh."

This is the arcane way. It is high mysticism, whether knowingly practiced, as science, or unwittingly and spontaneously exercised, as inspiration. By science, which is the knowledge of invariable orderly processes, inspiration follows speedily. By inspiration, to which great works are easy and masterful deeds are simple, the science comes slowly following after.

The mystics of all ages have trusted to their inward eye. While turning it to behold their own personal emotions or affairs they have wrought out no beauty of action or quickening language. While directing it toward the unnameable and undescribable King of Kings, they have astonished their own age and all ages, by their miraculous performances and noble aphorisms.

What made the shoes wax not old upon the feet of the Israelites forty years in the wilderness? Their gaze was ever toward the High Imperishable One, and even their garments partook of His unspoilable beauty.

What saved Hezekiah from dying, when even the powerful Isaiah had declared, "Thou shalt die, and not live"? His outer eyes with their dimming sight were following the uplook of that sense which we are all in constant use of for life or for decay. So swiftly did the life river come rushing down that flume of immortality that even the death-dealing Isaiah felt it, and turned to cry, "Thou shalt live!" He must now speak in tune with Hezekiah's resistless vision, for in the pathway thereof there is no death.

What turned Jacob from destruction, when reasonable

terrors shook him all night long, by the Jabbok Brook?
"I have seen God face to face, and my life is preserved."

What lifted Job out of his boils, and set his feet in
joyous security, at a time when the children of fools and
base men held him in derision, and the days of affliction
had taken hold upon him? "My witness is in heaven,
and my record is on high." What saved Daniel from
the jaws of famishing lions, giving him answer to the
lamentable voice of the king, as from within a calm
tabernacle? He had watched the untrammeled God and
His fleet angels.

Turning his gaze from the faces of men to his own
divine ego, Julius Caesar wrought over his fellow-men
like a god. "It does amaze me," cried Cassius to Brutus,
"that a man of such feeble temper should so get the
start of this majestic world!" Though Caesar's gaze be
high, yet it is not to the Most High, the Hebrew Hosea
would explain. High mysticism calls for highest uplook
toward the glory of the Highest.

"Thou canst upraise thyself by thyself, and rouse
thyself by thyself; for self is the lord of self, self is the
refuge of self."—*The Bhikshu.*

Cervantes beheld with persistent inner eye the image
of his mad Don Quixote, and ended his life in a mad-
house. Why not, if "that thou seest, man, that too become
thou must"?

Elisha set his watchful eye toward the Cause of Elijah's
greatness, and not toward the prophet speeding starward
in the fiery chariot. "Where is the God of Elijah?" he
calls, and down over his fingers fall curative ethers that
change poisoned pottage to nutritious food; into his
breath runs quickening fire that the dead cannot resist;

salt takes on a new savor, bread and corn forget their limitations, at the new tones of his voice.

"He that looketh toward Me, though least among men, his words shall be regnant."

And also, "Me whoso worships, he, completely transcending the qualities, is able to become the Supreme."— *Bhagavad Gita.*

A meek man prayed, "Show me, then, O King of all those mystics of superhuman powers, Thy Exhaustless Self."

And the meek man cried, "I behold Thee! Thou art greater than Brahma! Thou art of infinite valor and immeasurable power! Thou art the Primeval God! Thou art the Knower! There is none equal to Thee! O Thou with majesty unimaged! I behold Thee on all sides!"

By vision toward Transcendence the meek man became awake to Immanence. Omnipresence is but the garment of the Highest. None can find the Tao by way of discoursing of Omnipresence, Omnipotence, Omniscience. By the uplift of the inner eye toward the countenance of Him that weareth these garments, the two outer eyes are baptized with high altar fires to see the glowing land of splendor through which we ever walk, the finished work of One who saith, "Behold Who hath created."

As the mystically opened eyes behold the everywhere-completed splendor, the shadows of disorder are not remembered. But this glowing land yields not its sights to him whose mystic eye has not brought back over the pathway of obedience to the Heavenly Edict the soft alkahests that dissolve the films of blindness.

Ideal philosophy strikes the lofty note of the Sacred Edict when it forgets to maunder and prowl among reasonings begotten of unlifted vision. Now and then its voice rises like the sound of an invisible choir on the airs of night: "Keep your eye on the Eternal and your intellect will grow. Honor and fortune exists for him who remembers that he is in the presence of the High Cause."

The Egyptian thrice-great priests of Amen Ra caught the soundless teachings of the heights: "He is by Himself, yet it is to Him that everything owes existence. Becoming eye-witnesses, behold Him, and in beholding be blest. He is not light, but the Cause that light is. He is not mind, but the Cause that mind is. Nor spirit, but the Cause that spirit is. Let us lay hold of the Beginning, and we shall make way with quickness through everything. For the spectacle hath something peculiar; those that shall attain to the contemplation, it detains and attracts as the magnet stone the iron. But now as yet we are not intent upon the vision. So many men are body devotees they can never behold the Vision of the Beautiful. Why, O men, have ye given yourselves over to death, having power to partake of immortality?"

The Chinese of old had sages who spoke of returning to the High Deliverer:

"Returning to the Root means rest. He who regulates his attitude by Him will become one with Him. He is the good man's treasure and the bad man's deliverer. If princes knew the Tao the ten thousand things would of themselves reform. They would be restrained by the simplicity of the Ineffable. Homeward is the Tao's course. Who knows the way that is not trodden, and the argument that needs no words?"

The Hindu watchers toward the fronting horizon sometimes lifted their forward-caught, kundalini-bound sight, to the topless Heights, and hymned the rise of man from death and reincarnation: "Whoso worships **Me**, committing to Me all actions, regarding Me as the Supreme End, and to nothing else turning, for him I become without delay, the rescuer from the ocean of death-bearing, migratory existence. By reason of My being the Onlooker the universe revolves. Those devoted to the gods go to the gods; to the ancestors go those devoted to ancestors. Those go to the evil spirits who worship them, and My worshipper also comes to Me. I am beyond the destructible, and superior even to the indestructible; therefore in the Vedas am I called The Supreme. Whoso sees the Supreme, sees indeed."

The ancient Hebrews filled their scrolls with prophecies of the day when all mankind should look to the far heights for the opening of their outer eyes to see the supernal lands through which, ever stumbling, they with downcast gazing do daily travel. And the pages of their sacred books blaze with inspired urgings to greet the on-looking Deity: "The eyes of man, as of all the tribes of Israel, shall be toward the Lord. And the Lord shall be seen over them. The Lord of hosts shall defend them." "Say unto the cities of Judah, Behold your God!" "Seek ye me, and ye shall live."

The Taoist declared that this is the rest for which the earth-wearied are panting. The prophet of Israel saw that the coming rest from competition and struggle would be irksome to the age of hurry: "The burden shall be rest, in the day when the eyes of man shall be turned toward the Lord." "This people refuseth the waters of Shiloh (rest) that go softly, and rejoice in Rezin" (warfare—strenuous exertion).

The saving effects of the exalted attention are oft-times proclaimed by the Jewish psalmists: "Because thou hast made the . . . Most High, thy habitation; there shall no evil befall thee, neither shall any plague come nigh thy dwelling." "Who is like unto Him who exalteth Himself to dwell on high?—He raiseth up the poor out of the dust—that He may set him with princes."

History discloses that no word of self-disparagement or thought of fear counts against the saving grace that hastens to defend, or against the tender mercy that upholds, when that deathless soul faculty, the inner eye, lifts toward the Absolute beyond the Light, where not Spirit, but the Cause that Spirit is, doth ever call, "Behold who hath created."

"We have no might against this great company that cometh against us; neither know we what to do: but our eyes are upon Thee." And the Ammonites, Moabites, and Seirs, or the difficulties, inherited difficulties, and causes for discouragement, fled away from the besieged Jehosophat.

In simple meekness the king had stated his humiliating status, but he did the one thing he with all his army knew how to do—he looked, not with some mysterious sense we have to search for, we who are commanded to lift up our attention to the same all-accomplishing One, but with the everyday-used inner sense with which we can look back to our native city, or forward to the sunset.

This subtle faculty, swifter than the fleetest thought, being steadfastly rested upon any unknown point, can bring back to the waiting mind all the facts that pertain to the resting place. That we have let it fall most abidingly upon already transpired events, and drawn it away from the unexperienced, has been our own choice, indicating not at all the inadequacy of the able sense.

Columbus set his eye toward an unknown and un-believed-in shore, and landed his ships upon it. Thus goes he toward the unknown I AM, who sets his eye Himward. "Take sanctuary with Him alone, O Bharata's son, and thou shalt find the eternal abode."

Speech follows the direction of the visional sense. A man's words therefore soon expose why he is unfortunate or triumphant, great or inconsequent.

"Therefore will I direct my prayer unto thee, whom my outer eyes behold not, and I will look up. Early in the morning will I lift mine eyes unto thee."

"The ruffian looked at me, and wrought against me . . . diseases. So mayest Thou heal me, Thou most glorious One."—*Zend Avesta.*

It is the lifting up of this sense out of the network of materiality, the wheel of incessant grind, that takes man above his disasters and difficulties. "Mine eyes are ever toward the Lord; for he shall pluck my feet out of the net," cried David.

David's net was the wheel of events that harassed him, exactly as untoward events and disappointing circumstances worry the sons of men today. Down into these shadows streams a divine radiance, discovering to such as turn their gaze toward the Source of the Light Hypostatic, another outlook over affairs, more than compensating for the failures that menaced while the gaze was buried in misfortunes.

One above looketh toward man and his affairs. He is of purer eyes than to behold evil. Looking unto Him giveth some gleams of His view, for, "In thy light shall we see light."

Nothing David could say disturbed these effulgent beams from doing according to their own upholding ordinance, when Ezra was fearing his downfall. "When I said, my foot slippeth, thy mercy, O Lord, held me up," he gratefully acknowledged. Had he not practiced the precept of the sages of the ages, by which practice he must experience that they who look to the far Heights never falter? Had he not looked to the Source of the mercy that saves? Notice how the Greeks and Romans thanked the merciful beams from heavenly Mercury, touching them with magistral to poverty, and removing from their heads the guilt of their deceits.

Innocentius of Carthage, overcome with speechless emotions of fear and grief, looks to Him who alone can strengthen him for his crucial hour. Suddenly he finds that the surgical operation he has prayed for strength to survive, has been performed by invisible agency, and the saws and knives of material science are not necessary. For the angel of the Lord encampeth round about them that have the single eye that filleth the body with light, and delivereth them.

"Thus shall all the bodily world become free from old age and death, from corruption and decay, forever and ever."—*Vendidad.*

Ignorance counts nothing against one whose attention is steadfastly set toward the Countenance that shineth as the sun in his strength. Each one of us is darkly untutored on some vital point. In the day of effulgence from above, the ignorant master and the ignorant scholar shall perish out of the earth. For they shall all be taught of the High Supreme, not wisdom, but the Cause that wisdom is: "Thus will I magnify myself, and I will be known in the eyes of many nations." "And I will show thee great and mighty things, which thou knowest not."

Of new information has the Original of wisdom abundant store, to give in liberal measure when He is sought as the Author of intelligence. Therefore exalt Him and He shall shed new light upon thee, and upon all the inhabitants of earth. For by the obedience of one shall many shine forth.

Speak unto Him face to face, and no longer speak of Him. Speak unto Him over and over, as did Asaph the seer. Three times in the midst of his song did he chant, "Turn us again, O God, and cause Thy face to shine; and we shall be saved." By repetition he welded the attention of his wandering-eyed, weak-minded people toward the saving and illuminating heights.

No sage of earth has ever declared himself any other than a *seeker* after the way of the light that can raise the dead and heal the foolish; but Jesus of Nazareth said, "I *am* the way." Appolonius, who cured the diseased and called back the dying, travelled far to find if Indian or Egyptian priests could give him the law of life. But none could declare it, for all that they had spoken of the life-bringing light had been spoken in moments transcending their natural reason. "I *am* the life," said Jesus of Nazareth.

Gautama, who wrought many miracles, proclaimed himself a seeker after truth. "I *am* the truth," said Jesus of Nazareth. "We look for one to overcome nature's dominion," said Plato. "I have overcome the world," said Jesus of Nazareth. "I know that Messias cometh, which is called Christ: when he is come, he will tell us all things," said the woman at the well, echoing Plato's expectation. "I that speak unto thee am *he*," said Jesus of Nazareth. "I know that my brother shall rise again in the resurrection, at the last day," said Martha. "I *am* the resurrection," said Jesus of Nazareth.

This Man demonstrated His declarations by prompt proofs. He set the bands of death at naught, saying, "No man taketh my life from me, I lay it down myself." He nullified the limitations of matter, as, looking up, He multiplied food, and walked upon the waters.

And whether this Man is speaking as an historic character, not yet having shown that in His own person He transcends death, or as a risen and triumphant glory, exhibiting to all beholders a body that cannot be absorbed into death, He is ever setting His seal upon the doctrine that had preceded Him, that all great transactions come into manifestation by reason of the right view of some steadfast seer.

When disasters of nations come, and earthquakes, with seas and waves roaring, "then look up."—Luke 21.— "And Jesus looking up, . . . cried with a loud voice, Lazarus come forth," and the dead arose.—St. John 11.

On the three circuits where He found mankind struggling, He met them with the reviving elixirs of the heavenly vision, and caused them to outdo themselves. He put into living text the lost old Persian declaration that, "with right glance and right speech a man superintendeth the animate and inanimate."

On the first circuit, He stretches out His hand and touches wine, bread and clay, and they obey His will to step out of their captivity to habit. The wines of the mystic islands rise through the Cana waters. Bread unfolds from the ether's mysterious opulence. Clay hides the sightless eyeballs till the eye divine sends healing light, and clay shows strange hidden fire as the child of Nain quickens to life.

On the second circuit, He sends forth His voice and there is overplus of increase for the needy, and His hearers learn the mystery of the Logos, alive in every spoken word baptized by beams from the life-giving God.

On the third circuit, He warms the fishermen with coals which no man's hands have kindled, and prepares them to live henceforth by the dispensation of daily miracles wrought from above, that they may be the joy and enlightenment of ages to come.

On the first circuit, He finds people appreciating the tangible and material things of life, and He blesses the material things with something from above, but He says, "Flesh profiteth nothing."

On the second circuit, He finds certain among His hearers advocating the power of thought, urging the dominion of mind, and he blesses the thoughts of mind with something from above, as He says, "I will give you a mouth and wisdom," but also says, "In such an hour as ye think not," and "Take no thought."

All the transforming power which He uses on matter and mind He draws from above, teaching plainly that matter and mind must forever keep within restricted bounds of performance, till all the world looks up and draws down authority to unseal their limitations. "Canst thou by taking thought (alone) make one hair white or black?" "Blessed are those servants whom the Lord when he cometh shall find watching."

Those who set their attention toward the Countenance of the High and Lofty One inhabiting Eternity, are in the way of those ransomed from sin, disorder and death. And the ransomed are offered two songs: "The Song of Moses, and the Song of the Lamb."

A song is a perpetually recurring note of speech or singing, concerning some one theme. "I am become their song," cried Job. The ransomed return with singing. They know the Name of the Highest, which stands among men for The Absolute, as Origin of Being, Might, Majesty. . This Name was the song of Moses and of Zoroaster, those personifications of strength in leadership by the inspiration of Deity.

It is the Name taken up by all who lift the incorporeal eye toward the Author of Being, Might, Majesty. It is the Name the earliest known Egyptians had buried with them in their tombs, as full of the significance of immortality. It is as immaculate as the vision that is uplifted.

It is not the final name of the Cause of Being, Cause of Truth, Cause of Spirit; for as to proper name for the Father, the Unbegotten, there is as yet none known among men.

"These terms—Father, God, Creator, Lord—are not names, but terms of address derived from His benefits and works."—*Justin Martyr.*

But the Name which is called the Song of Moses is the highest name speakable by man at his present stage of expression. It has no reference to benefits or works. It stands by itself alone. It is applied to no other but One. It is, I AM THAT I AM. The term of address, or name God, stands for many objects of worship; the substantive Spirit, has many significations. It may mean one of twenty different descriptives; the name Lord, is employed in ten different ways, but the I AM THAT I AM is One. "When the children of Israel shall say to me, What is His name? what shall I say unto them?" "Say . . . I AM hath sent me unto you." And He "led them by the right hand of Moses with his glorious arm, dividing the water before them, to make Himself an everlasting Name."

The Name I AM, addressed to the Highest, wakens the spirit of authority, majesty, undefeatable courage, in the breast of even the meekest and weakest of men. "I have wrought with you for my name's sake," spake His voice to Ezekiel.

The Name I AM THAT I AM brings up from the deep wells of hidden strength in all men the sincerity, boldness and intelligence of leadership, and that originality of action and language which have characterized the heroes of the ages, whose names have lived so long in history that they have become myths.

It is recorded of one of these, that in deepest humility, asking of the Self-Existent face to face, His most order-bringing Name, he heard the words, Ahmi Yot Ahmi— I AM THAT I AM. And this man became ruler of a kingdom, and founder of the Wisdom of the Magi. He had touched the leading note of that Ineffable Name which is key to the mysteries of the universe.

This Name is the first utterance of those who set their attention toward the Heights, whence fall the kindling sparks that burn away the films hiding the finished splendor of the realm through which we walk.

And the Song of the Lamb is the second utterance of the upward-visioned among us. It is the name JESUS CHRIST. "In My Name," said He that was slain. "In His Name," said His disciples. And it is declared that they never preached any doctrine except the power of His Name. This was their Song. It is a Name as immaculate as the Name I AM. It always means, God with us. It is the Amita Buddha, the Ahura Mazda, the

Emmanuel. It is that Name of the Lofty and Everlasting I AM which represents His nearness and immanence. Name above principalities and powers, it is the
Name of newness, of healing, and of comforting tenderness. It gives the baptism of the quickening Spirit. It
is the greatest and quickest God-formulating Name.

It is the Name that restores the Lost Word, the now
unspeakable Name of the Self-Existent Deity.

The Moravians hymn the power of this Name:

> "Should I reach my dying hour,
>   Only let them speak that Name;
> By its all-prevailing power
>   Back my voice returns again."

And they tell of miracles of calling back from the
dark defile of voiceless death to sunlit life, by the resur-
recting energy of this Name.

The rulers of the Jews in Jerusalem, Annas the
high priest, and Caiaphas, and John, and Alexander,
and many that were of the kindred of the high priest,
A.D. 33, knew well the Magian power contained in certain names, and they asked, "By what name have you
wrought this miracle?" "By the Name of Jesus Christ,"
answered the Christian Apostles.

The risen Christ, appearing suddenly, said, "Preach
repentance . . . in My Name . . . beginning at Jerusalem." And Jerusalem means THE SELF.

Begin with yourself to repent, to return. Lift up the
willing inner sight toward the Supreme One, whose
Soundless Edict through the ages is, "Look unto Me,
and be ye saved." Taste the first manna which the
upward watch sprinkles over the unfed brain and heart.
This is reasonable service. It is mirific obedience.

Facing toward the Heights, where the smile of the Comforting One begins its beaming Omnipresence, Omnipotence, Omniscience, speak from the heart the two greatest Names ever written or spoken on earth. They are the only response the heart can make when the mystic eye is first uplifted. Without the uplift of the deathless sense the Names may be but heathen repetitions. For liberation is not achieved by the pronunciation of the Name without direct perception. But consonant with the upward watch, these terms of address to Deity are the planting of the feet upon the rock of power and the transmeable hills of security.

"He sent from above, he took me; he drew me out of many waters; . . . Thou also hast lifted me up on high, above them that rose up against me."

Whatever comes upon you this day, or threatens to disturb or overthrow at any time, turn then from it toward that High Deliverer looking hitherward, and within the silent heart, sing the two Wonderful Songs of the Seers of the ages:

"O High and Lofty One inhabiting Eternity! Clothing Thyself with Thine own Omnipresence, Omnipotence, Omniscience, as with a garment—hiding Thy goodness and majesty with names, and unspeakable names! I know Thou Art, and the Name of power and glory I must address to Thee is, I AM.

O Countenance beholding me, looking toward me through the ages! Breath of the everlasting life in me, and manna to my fadeless substance! Thy Name that folds me round with tenderness, and lifts me high above the pitfalls of my human destiny, is, JESUS CHRIST."

The Practice of the Presence of Deity, through obeying His one great commandment, "Look unto Me," removes the sense of limitation and danger. This Second Study tells us to persist in obedience till we thoroughly experience the five liberations mentioned on page 32.

II

REMISSION

We are so constituted that when we are told that the Divine Edict is, "Look unto Me," we lift our inner visional sense to look toward the High Cause. The Spanish mystics urged mankind to look a thousand times an hour toward the Vast, Vast Countenance that shineth as the sun in his strength.

The beams of that Countenance are hot with healing. Proculus the slave felt them melting both his sickness and his chains, and Severus the Roman Emperor chose Proculus to live in the palace with him, because an ever-falling grace from above undid the disorders of those who came near Proculus.

Something is ever gently wooing us. It is the sin-undoing Saving Grace. It rides swiftly to our freedom on the thrill of our recognition. It exposes the Elysium which the saints of old told of in song and sermon. Every beam of light and every waft of air from sighting toward the Unsullied Heights is the touch of the dissolving alkahest, the remitting mystery, the saving grace, removing some suffering, exposing some joy. Is it not written, "Look unto Me, and be ye saved"?

One who had been utterly set free from the might of the flesh and its death, rose, an untrammeled Being, and said, "Preach Remission." Preach the removal, the putting away of the consequences of the downward vision, which appear as evil, matter, lack, pain, decay. Preach the freedom of those who notice that Deity onlooketh them.

A principle is a comprehensive law. As Jeremiah was brought to sickness and martyrdom by gazing much toward affliction, so they who look to the Unweighted First Cause are unweighted of sickness and the possibility of martyrdom. "Behold, I will . . . lay thy foundations with sapphires"—liberty. "They shall fight against thee; but they shall not prevail against thee."

The mind is not capable of bringing anything to pass except it be transfixed by inward visioning. Inner vision is the vital essential to the mind. When this faculty is exalted, the mind quickens with original ideas and has high instructions.

Hegel, turning his attention toward causes, got deeper than thinking, and wrote in his "Introduction to Logic" that we secretly perceive *toward* an object before thinking it, and it is only by having constant recourse by inward viewing that then the mind goes on to know and comprehend.

The High and Lofty One that inhabiteth Eternity offers to undo our weakness and our wretchedness, the laws of matter and the veneers of time, if we, undiverted, seek His face. He offers a new language and the end of the world: "Look up to the fields white for the harvest." "The harvest is the end of the world." "They shall speak with new tongues."

As we are like those we face, dropping their unlikeness, it is not strange to find the mystics of all times joyously exclaiming that the Undifferentiated Self-Existent, the Abyssmal Naught, has remitted for them the five dark unlikenesses to Himself into which their aberrated watch had warped them. They have not been seeking liberation from bondage, they have only been seeking His face, according to the Sacred Edict, "Look unto Me," yet liberation has been as complete through the ecstatic moments of their contemplation as if they had entered Paradise. Read how the cosmic laws of matter and mind let go for Parmenides while he was seeking the High First Cause, "not life but the Cause that life is."

The experiences of the mystics have been reported as the shouts of the free. Their shouts have been the creedal formulas of philosophical and religious organizations without number. The zealous lovers of the formulas have often forgotten that matter does not loose the grip of its law if the vision is not toward the heights; that evil lets go its claims only when the Dayspring from on high drops its dewy sunshine into the heart; but all the same the world has loved to hear the lovers of the formulas sing the non-estness of evil, matter, pain, decay. The world has always sought its mystic nihilists when it has wanted its prison doors unlocked: "I will turn back your captivity before your eyes, saith the Lord."

High mysticism is divine nihilism. Truly, there is no knowledge except what is taught straight from Him who saith, "I will instruct thee, and teach thee." Truly, there is safety from drowning for any Peter who looks away from the stormy waters of human existence to Him who saith, "The flames shall not kindle, the waters shall not overflow, . . ." ". . . nothing shall by any means hurt you."

It is preaching remission when we tell the Unweighted Light face to face, that we know our surety of unburdened life under His healing smile. It is a prayer; and prayer is ever a psalm of freedom: "Am I not an apostle? Am I not free?"

If the vision be on high, where the illimitable skies in untrammeled buoyancy shine chastely down, purity of moral tone and flawlessness of body are manifest. Vanity, dissembling, cowardice, are remitted, dissolved; sickness, weakness, disease, are removed. The original Self, denuded of its crass teguments, forgets its history in matter. The former earth is forgotten. It does not come into mind any more.

He unto whose face we look hath vouchsafed to no man His Name, and none as yet knoweth His nature. We know His promises, His gifts, His responses; but as our only carrying energy, the inward visional sense, has been engaged in fetching either gladness or sorrow from other objectives than His glory, we have nothing sublimer than Life, Love, and Spirit, of which He is the Giver, to describe. These being His gifts, and already among the tangible and practical experiences of man, it is not surprising that the subtler triumphs undergirding those who have sought the Giver, and not the gifts, have not been understood. For only the Original of knowing can say, "I will show thee great and mighty things which thou knowest not."

Under the worship of Amen, the Unknown and Unknowable High Cause, Thebes, the metropolis of Egypt was the seat of kings, and triumphed over all the world. Under the worship of Aton—Life, Truth, and Love, or Truth, Happiness and Sunshine, Thebes flourished in splendor for a period, was rich, magnificent, and pompous, and then suddenly went out. Life and Love may not say, "Look unto Me and be ye masters of life." They may not go higher than repletion with their like. Only their Author is their Master and He only can confer mastership. "All the nations . . . shall fear and tremble for all the goodness and for all the prosperity that I procure . . . thee, saith the Lord." "Thou canst not behold Me with thy two outer eyes; I have given thee an eye divine."

"Touching the Almighty, we cannot find him out . . ." "with God is terrible majesty." "Behold God is great, and we know him not," chanted the awe-struck Elihu.

Paul, in his Mars Hill address, declared the Unknown God who giveth life and spirit. He could not describe His nature nor name His greater Name than I AM THAT I AM. But it was immortalizing to Paul himself to call the gaze higher than life and spirit, to the Ain Soph of the Cabala, the "Great Countenance of the Absolute, above thinking and above being."

"My Father is greater than I, was the upward-calling statement of Jesus; and His eye being ever toward the Divine Original, He was ever Master of life. "I lay down my life, . . . and I have power to take it again," He said.

As there is rich newness over the Tao, or Highway, we need not fear to let go all that goes under its dissolving beams. Is it not written that no man can see that Face and live according to his former estate! That his former estate shall shuffle off—be remitted?

Sell all, let all move aside—let go, and give to the Poor, the Unknowable Absolute, the Unhindered God, the Unweighted I AM, the Predicateless Being. This is the Universal insistence of inspired mystics. We have only one thing to give, namely, our attention. There is one Poor, namely, the Unencumbered First Cause. "Who holdeth fast to the High First Cause, of him the world shall come in quest."

Preach the deliverance of the captive. Acknowledge high. Tell the One High Cause, that being Untrammeled Freedom in Himself, all who look to Him are untrammeled. Tell Him that fear and doubt depart. Tell Him that captivity itself is led captive, and only un-vanquished Soul salutes Him.

The dissolving alkahest, the gentle grace that falls down over the track of the vision, has been praised by the sages of the ages, for there is surcease of world pain in its white softness: Behold the gentle Neutral that taketh away the mistakes of the World.

Five grievous shades slip off the earth. They are the foolish virgins with no oil of healing and no oil of il-luminating in their most eloquent declarations. No one describing them was ever to himself or to his neighbors, the oil of joy for mourning, nor ever the inspiration of wisdom while detailing their processes. It is the passing of these five shadows that has caused the five great shouts of liberty, the high sounding Pehlevi, the psalms of re-mission, the prayers of the released:

1. Steadfastly facing Thee, there is no
   evil on my pathway.
2. Steadfastly facing Thee, there is no
   matter with its laws.
3. Steadfastly facing Thee, there is no
   loss, no lack, no absence, no
   deprivation.
4. Steadfastly facing Thee, there is
   nothing to fear, for there shall
   be no power to hurt.
5. Steadfastly facing Thee, there is
   neither sin, nor sickness, nor death.

Preach remission, said the Risen Christ. Preach that the iron gates open of their own accord for upward-gazing Peter.

Preach that the stone of interference looming on our life path is rolled away, as for the two Marys.

Preach that palsy falls off Æneas, and death falls off Dorcas. Understand what the Vedas are hymning:

"O thou Unshaken One!"
By thy favor my delusions are destroyed!"

Matter and its laws of mind are the fictitious generations of ofttime downward glancing with our efficient visional sense. When this sense is lifted up, what seemed external exists no more at all. The inner vision leads off the other senses and if it is exalted toward the Healing Onlooker all the senses aver health. "For I am the Lord that healeth thee." "The way of life is above to the wise, that he may depart from hell beneath." "Seek ye my face and live."

It is a very subtle doctrine that man is like that to which his inner eye is oftenest directed. It has been called the secret doctrine, because whoever discourses on the laws of mind, or describes the omnipresence of Life, Truth, Spirit, has not touched the secret of Deity's look toward him and his look toward Deity. In this sight, or science, is denuding even of Spirit. "Blessed are the poor in Spirit." Something transcending Spirit smiles. Let the Spirit blow where it listeth. "There is no man that hath power over the Spirit," said Solomon, for Spirit is the servant of the High Deliverer—the I AM THAT I AM. "Behold, I will pour out my Spirit unto you."

After years of austerity and singleness of eye, the Hindu mystic finds that the Deity who looketh toward us, saying, "Look unto Me," is not Spirit, for Spirit arises in opposition to matter, and the Deity is above distinction. He notes that he only who perceives the Lord as Differenceless, goes to the Supreme End.

"He is not Being," says Erigena, the Irish mystic, "for there is an arising of contradistinction; He cannot be called goodness, for goodness is opposed to badness, and God is above this distinction."

Proculus, learned in the ritual of the world's invocation, concludes that Deity is best described by negations, since only His gifts are knowable. "He is not that," we may insist to all men's descriptions of Deity. Job finds his swollen flesh dissolving, when he sets his witness in the heavens: "Thou dissolvest my substance," he cries. Then the strong man of him springs up under the remitting but energizing light, and his last days are more triumphant than his youth and prime.

Specialists multiply that which they investigate. There shall never be an understanding of how fadeless health is roused, so long as the physical system, that faithful register of woe and vigor brought on the wings of secret viewing, is sought as the informer. Only by the study of the Uncontaminated One that inhabiteth Eternity shall unspoilable wholesomeness laugh in the substance of living creatures. Only His way upon the earth is the saving health of the nations.

The material body is a hard taskmaster. What it ought to be fed with, and how it should be housed and trained—see how it worries us with never telling us. The sons of the Tao know that neither if they eat are they the better, nor if they eat not are they the worse. Every mouthful shines with new mystery, and buoys up the system as on wings of might; every abstinence leaves the veins free for the sunshine of the Beatific Uplands to flow in radiant strength along. Who can prove this till he has ofttimes torn his gaze from the wheel of things to behold the Unencumbered Highest?

The mind is a wearisome objective. Its thoughts have laid claim to great powers of destruction and great powers of building. With the brightness of the I AM beaming upon them, even their wrath is praise of the Unthinkable Absolute. Jesus can look around with anger, being grieved and the withered arm stretches forth restored whole as the other. Under the baptism of the Divine Smile, the wickedness of the wicked shall not destroy, and the righteousness of the righteous shall not save. The Tender Mercy, the Dayspring from on high, remits the thoughts of the mind. Man's inheritance of things that have not been conceived by mind comes into sight by looking to the Unknowable, who originates new knowing. "I will teach thee." "I will turn to the people a pure language."

He who watches for the erroneous thought that caused the malady of his neighbor shall find it alighting upon himself: "The watchers for iniquity shall be cut off; that make a man an offender for a word."

"We wrestle against the rulers of the darkness of this world," said Paul. He is speaking of the wrestle of the morning of health that dawns on the upward watch, with the midnight of disease that glooms with looking toward mind and body. All the ways of darkness are removed by the Light that falls with remitting grace upon him who notices that the Deity looketh upon him. "Look up to the fields white for the harvest." "The harvest is the end of the world." "Speak ye comfortably to Jerusalem. Tell her that her warfare is accomplished." There is one universal solvent. It is the falling alkahest, the whiteness that makes death let go, that looses the bands of palsy and of pain. There is no pain facing Thee.

Steadfastly looking for high news, Origen finds that "evil has no substance." Plotinus, on the same quest, becomes aware that "matter is nothing." Hezekiah, more awake than they, rises in free majesty, because, "The earth and all the inhabitants thereof are dissolved." Isaiah, greatest of the seers, finds that "All nations before him are as nothing, and they are counted unto him as less than nothing." Job, meekest of all under the ever-beholding Solvent, yields himself in joyous dissolution—"Thine eyes are upon me, and I am not."

The assurances of the stately scholars called mystics, have ever been, that all the hurting powers are nullified—remitted, for him who looks away to the Divine Original:

"No weapon formed against thee shall prosper."

"No bad fame can hurt thee."

"Thou shalt be far from oppression."

"Terror shalt not come near thee."

Their accepted formula has not been that the world is divine, and all things are God, but the world is nothing—the Lofty One inhabiting Eternity is The Alone—The All.

Looking downward, we weep at loss and lack, while the offer has ever been that there shall be no lack for the beholders of the smile of the Ain Soph. Obeying the Sacred Edict, "Look unto Me," woman's cry of no wine of life—the health, strength, praise, of which she never feels she has a plenitude—ceases. She hungers no more. She wants for no good thing. "Woman, what have I to do with thee?" says Jesus, when Mary declares, "They have no wine." Woman has ever been the propagator of lack. She is to be first to declare, "They shall want for no good thing facing Thee."

Man's strenuousness on every line desists; he labors not, he takes no thought. There is a way that looking toward labor and lack has hidden. Looking to the Heights, away from labor and lack, the way is visible. "I will lead thee by a way that thou hast not known." "The people shall not say, I am sick." "They shall not see death." "I am the Lord that healeth thee."

This truth of the nonhindrance of matter and mind, as the beams of the shining Countenance penetrate to the hidden man of the heart, has the testimony of many sages of the ages. It is the mystical vision and union which Dionysius discovered gave him "that most divine knowledge of Almighty God, which is known through not knowing."

It constitutes that union which Plotinus declared, "Enkindles our life flame, giving rest to the soul now fled up, away from evil, to the place free from evils." How worth while to view above time and sense, preaching the inevitable remission, taking the instructions of the sages of the ages who have experienced it.

Sometimes these great forerunners have called our mental, moral and physical characteristics, and comports, "our garments folded around in our descent to view the not-God." And they show how one by one, these garments have fallen from them on their upward-fleeting vision. "Love of honor is the last garment to be stripped away, as we show ourselves more like the Divine," is remission preached by Proculus.

According to all these illumined ones, it is the Omnipotence through all things that binds them all in such sympathy. The crawling worm is brother to the archangel, in the fact of his central spark being God. And wherever remission is experienced, there is the miracle of the creature divinely transcending environment. Preaching remission uncovers the divinity at the center, because it entices the eye heavenward, whence the uncovering daysprings hail.

The illuminati of the world, each manifesting according to his own recognition of the Supreme, have been at all times living proofs of the efficiency of setting the watch toward the Self-Existent Heights. Unto some it is the manifest readjustment of environment. Unto others it is the forgetfulness of environments. Unto some it is joy in shining health of body. Unto others it is forgetfulness that the body exists.

Union with divine Freedom, the heavenly Poor, is fraught with resultants near and far. The world receives a treatment as the pioneer on high plains unifies with the free Light. For the watcher shines forth as the sun with the healing glory of his Father above. Why may not the dead rouse past their cerements, if some pioneer abides undiverted in the sight of that peace which Cosimo de Medici saw folding round Antonino, to such effect that it stopped earthquakes? The world awaits the great Peace Treatment.

Whether the Unweighted Heights are sought in coldly scientific mood, or in religious warmth, to inspire the particular from the Universal, that which would hedge the free Self removes. "Watch the Way," said Nahum, "so fortifying thy powers mightily."

Ezra is scientific: "I will lift up mine eyes to the hills, from whence cometh my help;" but he makes haste to be religious: "I have gone astray like a lost sheep; seek thy servant, Lord." "Great peace have they which love thy law, and nothing shall offend them."

He who learns that gazing upward toward his Father's

face is a liberating act, rouses with fresh hope. He senses that he is greater than what has heretofore happened to him. He forgets his calamities. He is on the pathway of salvation from the causes of calamity.

The free, wise, immortal center of man is the begotten of God. Only this uninjurable and shining principle is offspring of I AM THAT I AM. Not only the free and unspoilable soul, spirit of Jesus, but the soul, the hidden spark, of Nero, is Son of the Highest; Jesus, being unclothed of matter with its mind and temper, Nero, being heavily garmented therewith. The upward vision saves Nero or Jesus. There is no respect of persons on the high watch. "Mine eyes are ever toward the Lord; for he shall pluck my feet out of the net," may be proclaimed by bad and good alike. "God is the bad man's deliverer," gratefully sang the ancient Chinese sages.

The remission loved by Proculus was the disappearance of memory and idea while raising eyes to the Predicateless One. The remission loved by those who lift their eyes to the Predicateless One in this age, is the removal of the hurting powers of life and death, riches and poverty, sin and virtue. They look for the day of Wisdom to break, and the shadowy night of ignorance to flee away.

The ordinance of the Highest, "Look unto Me," requires an individual practice. It compels a life of its own. It exposes the doctrine known to Jesus of Nazareth, who said, "If any man will do His will, he shall know of the doctrine." At the point of knowing right doctrine, an influence emanates which clarifies all atmospheres. The neighbors of the knower drop their errors of thought and conduct. They start to seek the highest good at the highest fountain. This is that cognition which irradiates, till kings consider that which they have not been told, as Isaiah prophesied.

Gazing toward the One and Indivisible, Parmenides learned that the phenomenal world with its origination and decrease, multiplicity and diversity, is nonexistent and illusion. This knowledge extended forth from him, rousing sublimity of character and conduct in his adherents. Leading a noble life was denominated the Parmenidian life among the Greeks for centuries, so influential had been the remission preached by Parmenides.

But even indeed were there no radiance from the watcher's clarified being, he would love and preach the High Eternal I AM, whose assurance is union by vision. The heart would praise and extol Him whose look toward man remits all unlike His own nature. To the real heart there is joy in the divine fact that there are treasures of knowledge laid up for him whose true foundation is freedom by the look of his God upon him, dissolving his mentals, and setting aside his bodily hindrances. To the heart there is gladness in knowing that remitted conditions leave exposed the secret original Self from which transcendent character springs forth, daily honoring the Father with the beauty of Soul Integrity.

The second angel sounds; man acknowledges the liberty that he senses by obeying the high mandate; and the mountain of all personal obligation rolls into the sea. The I AM THAT I AM, of Unspeakable Majesty, is seen to be the only Responsible One. Like St. Augustine, in his hurried glance, man attains to the vision of that which is: trouble as a shadow has flown; he cannot find it. Only untrammeled God is Real. "Is there a God beside me? . . . I know not any." Anxiety is no more.

With Maimonides, Spinoza, and Gerson, in their
diviner moments, proclaiming that evil has no existence
forever, the obedient watcher heavenward walks the
buoyant path of fearlessness. He bursts the bonds of
desire and its attainment. He breathes above ambition.
The wonderful God makes him a preacher of the heavenly remission. He tells of Him above Truth, whose
works are truth; of Him who saves from old age and
death, disease and poverty, ignorance and competition.
He joins the singers of the God-born Vedic hymn:

> "Destroyed is the knot in the heart;
> Removed are all doubts;
> Extinct are all the hidden longings,
> Upon beholding Thee."

-Let us set aside a day to telling over to Him unto
whose Divine Countenance we look, the wonderful re-
missions, the heavenly liberations promised to those who
ofttime turn away from smothering environments to
face the Lover inhabiting Eternity—the Lord of Hosts
His name. Let us boldly acknowledge, as we lift up
our eyes unto the Deliverer, the Limitless, "Because
Thou art the Unconditioned and Absolute, I also am
unconditioned and absolute. Because Thou art the
Free, I also am free. Because Thou art the Self-Existent,
I also am self-existent."

This Third Study is prepared so that even those who have not heard its subject matter orally can understand that the High Vision which awakens high thinking and incites to noble living has been the vital theme of the preceding studies. Something always antedates thought. That something is Vision.

"Look unto Me" is the Sacred Edict.

III

FOR-GIVENESS

When Aristides, wise Archon of Athens, was so ill that his physicians left him, he saw as in a half dream the goddess Minerva with her shield. She was to him in his dream even more beautiful than her Athenian statue by great Phideas. He called out to those about him to hear her words relating to honey from Mount Hythemus and the new diet he must observe. His family could not hear or see the goddess but they prepared the honey and arranged the diet exactly according to his report, and he soon recovered. Aristides declared that Æsculapius was with Minerva at the time, shedding healing breaths over all her balmy words.

From unprejudiced standpoint there does not seem to be any striking difference between the experience of the Archon of Athens attending to his two gods, and the Christians later on attending to their one God.

"When earthly helpers fail and comforts flee,
Help of the helpless, O remember me!"

Two gods with benignant smilings: one God with smiling benignance!

Throughout all time there has been tacit understanding that when half gods go the gods arrive. Some let go of their half gods with tears and lamentings, bemoaning the departure of all earthly helpers and hitherto comfortings; and some take hold of unseen help with

43

groanings and grim determinations, whereby they pain-fully earn their blessings, heroically forgetting, "Cast all your care . . ." "My yoke is easy and my burden is light."

It was easy for Aristides wise Archon of Athens to let go his earthly helpers for they all forsook him and fled, and it was easy for him to receive healings from his gods for he was half asleep when they dropped their sweetly worded balsams on his head. It is the wisdom of Jesus that He enjoins being wide awake and easily letting go, and wide awake and easily catching on, to the sprink-ling alkahests, all soundless nepenthe ever falling on all our heads: "Hurt not the oil of letting go; nor the wine of healing inspiration."

Repent, for remission, receiving the Holy Ghost that maketh whole.

John the Revelator is not choosing haphazard the chalcedony stone as symbol of the third lesson of divine law. The chalcedony signifies awakening strength: "Awake, awake, put on strength, O arm of the Lord." Look up to fields white for harvest; so shall old condi-tions dissolve; so shall the Holy Ghost arrive—white breath that maketh strengthening wholeness.

Mankind sticks to a triune of some kind. It is their mysterious instinct. Pythagoras called Three the num-ber of Divine Law. The Jews have always regarded Three as a specially complete and mystic number, and we may note that this *Study Three* with its wide appli-cation, holds all the twelve lessons in its norm, or pattern, if we read its purport aright. It is but interpreting the story of three-faced Hecate to our own generation. Whatever way men regarded Hecate, that way or face would she show to them.

So also does God the Father Almighty, miracle-working Jehovah-Triumphant to the great prophets: God the Son Almighty, miracle-working Jehovah-Triumphant to Apostolic Christians: God the Holy Ghost Almighty, miracle-working Jehovah-Triumphant to the Apostles of the Mystical Dispensation just winging its white influence across our awakening planet.

So far, this third dispensation of the Triune God in the Universe has not shown forth the mighty miracles of the first or prophetic Dispensation; nor did the second or Apostolic Dispensation show itself equal in grandeur of performance to the first; but the halt in splendor of achievement has had more universal promise in all that has been done, as if a whole globe were being bathed in softly stealing Brahmic Breath where astonishing whirlwinds had once glued the world's awe-struck attention.

When Moses, legislator of the Hebrew nation and founder of the Jewish religion, called to God, the Father Almighty, to divide the Red Sea before the two million Israelites fleeing from Egyptian bondage, he heard his God saying, "Wherefore criest thou unto Me? Speak unto the children of Israel that they go forward: and lift thou up thy rod, and stretch out thine hand over the sea, and divide it: and the children of Israel shall go on dry *ground* through the midst of the sea."

When Joshua needed that the light of day should keep on while the Amorites were fighting his people, till his people should win in the battle being waged against them, "Then spake Joshua to the Lord . . . and he (commanded) in the sight of Israel, Sun, stand thou still upon Gibeon," and the sun stood still till the Israelites had shown supernal fighting genius to the five kings of the Amorites with their combined armies.

When Elisha chose to open the eyes of his servant to see the Army of the Lord of Israel encamping in the mountain to defeat the army of the King of Syria, horses, chariots and solders, a great host, then his servant saw the Army of the Lord with their chariots of fire filling the mountain of Dothan, movelessly and soundlessly fighting for Israel till the Syrian hosts had no more power to hurt the army of Israel.

When Ezekiel, worshipper of the Father Almighty raised the dead, he raised a whole valley full.

When the Apostles, worshippers of Sonship Almighty, preached the Risen Christ Jesus, they raised dead Eutychus and cured many taken with palsy and lameness.

When worshippers of the Holy Ghost, whom the Father Almighty hath sent because of the name of the second dispensation, or the Jesus Christ Dispensation, do works, they do them on a small scale, but being determinedly related to Universal Spirit, or Brahmic Breath, which is Holy Ghost Influence they sweep the globe with inspiration. For now is come that Great Spirit, the wind in the wings of the messengers of Jehovah-Triumphant, and everywhere the sound as of a mighty wind from heaven.

Has it not been declared that a Breath of Brahma wafts through our common atmospheres, breathable by all who choose to inhale it as strengthening spirit?

> "O hither wafting breath of strength
> In Brahmic ether's keeping!
> Man may wax stronger day to day
> By the mystic way of reaping."

The Breath of Brahma was what Job was volitionally inhaling till it healed his mind of grieving and his bones of soreness. Inspiration, or inbreathing of the air-encompassed Ghost, or heavenly Breath, is sure healing of the mind of man; sure transforming of his thinkings; sure healing of his body throughout all its flesh and bones; sure healing of his affairs also; but who is found indrawing winds his nostrils take no note of, gazing towards a hither-wafting white breath his outer eyes see not? But truly, such only are those revivingly healthy among us who have learned that the healing breath never faileth, changeth never, abideth forever in miracle-working competence exactly as Job reported: "The breath of the Almighty giveth me life."

Why, O mankind, so decrepit, having power to partake of mystical renewal by quickening inspiration? "Ye shall receive power after that the Holy Ghost is come upon you"—quickening inspiration that waketh the hidden God-Seed; inspiration that cureth the mind of thinking; inspiration that draweth hitherward that Mind which no man as yet knoweth, the instantaneously working Ain Soph above thinking and above being: forgiving the mind of the world; hurrying along the promised speech that distills new health like morning dews.

With the sounding of the first angel, writes John the Revelator, there followed hail and fire, mingled with blood. The first angel is the first call to LOOK UP.

Obeying the call, the resistless rule of the skies with soft alkahests dissolves opposition. The strenuousness of the laborer and the anarchist lets go. The strain of human existence is hailed upon with blessings from above. A new fervour glows in the speech. The heart of man fires up with knowledge of a right, outshining the law of "Thou shalt not." The sense of plenitude wakes the shout, "All hail sweet riches!" for the groan "I want!" This is for-giveness. It is the beauty given for ashes that Isaiah saw across the ages.

In everything Hezekiah, King of Judah did, he was prospered, because his heart was in it. But though Amaziah did the thing that was right, there was no enchantment in his deeds because his heart was not enlisted. The fire that John saw mingled with hail is the kindled heart.

And the blood that commingled was the new type of man that is even now among us, healing and protecting when he appears, as the young man in the furnace protected the three friends of the compassionate king and presented them to him unharmed.

And one-third of the trees and all the green grass are burnt up by the fire of the new heart and speech, continues John. All the competitives that constitute the ginger and glow of human encounter, flourishing like trees for strength, give way to the miracle that sets each man into his right place. The hastening periods of childhood, youth, old age, forego. The man standing in the fiery furnace knows nothing of the seasons of life that are like the grass. Obedience to the mandate, "Look unto Me," introduces a new order.

"I will give power unto My two watchers," was the promise John heard from above, meant for any two of us who now begin the high watch that wooeth the God power. "These are the two olive trees." These are those enriched and set in authority from above, not by effort, not by worthiness, but by resistless grace falling over their high watch, as were Moses and Aaron of the upward-visioned Kohathites. "These are the two candlesticks," priests taught, "from above," speaking with new tongues, needing not that any man should teach them. These are the two olive branches, whose golden pipes send forth the golden oil of healing all unseen but resistless as the sky stones called healing hail; the golden oil of enwisdoming till kings know that which they have never read, and understand that which they have not been told; the golden oil of prospering till the poor lift up their heads with comfortings, wooed to comrade with those angels who minister cure to poverty. So shall the nations seek and find Ain Soph, the Great Countenance above thinking and above being.

With the sounding of the second angel, writes John, the Revelator, the sense of personal responsibility, of heavy obligation, rolls softly away. Look up whence the high laws hail, unburdening the tongue of talk of hardship. If the tongue is yet speaking of hardship its owner has not sensed the second angel's message.

With the sounding of the third angel, said John the Revelator, the star called "Wormwood"—ware-mood, the mind-preserving principle—wraps the conscious mind in sane security. The mind is forgiven its suggestibility to foolishness and ignorance. It is gathered to unsuggestible Illumination, *Jehovah-tsid-kenu*. Hitherto all healing has been directed for the body's benefit or for the lightening of hardship. Now it is that the mind no longer thinks, "I walk on the earth, or on the floor." The mind is cured of such thinking. It gives way to the sight of that foundation under our feet that is God Eternal. The mind is cured of thinking, "I put my head on my pillow." It gives way to seeing underneath the God arms everlasting. The mind no longer thinks "I breathe common atmospheres." It gives way to glad discovery,

"The Spirit of God is in my nostrils," and "the breath of the Almighty hath given me life."

We always become like those with whom we associate. Did not the observing Herbert Spencer conclude that man is more like the company he keeps than that from which he is descended? The youth, Evison, who associated with sellers of salves and plasters, never grew any eyeballs in his empty eye-sockets till he began to attend faith-cure meetings, where his attention was steadily urged toward the curing sunshine of the Vast, Vast Countenance ever beaming toward him. Iamblichus noted that certain men had taken on majesty and superhuman accomplishing powers from constant association with the powerful gods.

Associate with the alkahest and its remitting absoluteness is ours. Like Proculus at the king's court we bring dissolvings to pain. We associate by converse. "He that communicateth with me strengtheneth," saith the Lord of Strength. He that speaketh unto me wakeneth. He that toucheth me is cured. "To him that holdeth his conversation aright will I show salvation."

The third star was Wormwood, or the tonic of divine contagion. "Repent, for the remission, and ye shall feel the contagion," said Peter to the brethren. Repenting himself, he caught the curing contagions. "Now let signs and wonders be wrought!" he shouted. "By stretching forth thine hand, to heal, O God!" he said, as the curing flakes fell even upon his shadow.

"Hurt not the oil and the wine," cried the angel of the third seal. Hurt not the doctrine of denial, the cathartic oil of unburdening recognition and its speech, as, "Facing Thee, there is neither sickness nor death on my pathway." Hurt not the strengthening wine of affirmation, as, "there is none beside Thee. Thou hast forgiven my mind."

> "My bark is wafted from the strand
>   By breath divine.
> And on the helm there rests a hand
>   Other than mine."

"The third, the face of a lion," writes the entranced Ezekiel. The lion is emblem of strength, sovereignty, and princely achievement. We are as strong, as sovereign, as able, as our backing, our consort, our *aid-de-camp*. "I can kill a thousand snakes! I can build a house!" shouts the tiny child clinging to his father's hand. He feels the strength of his father. "By thee have I broken through a troop! And by my God have I leaped over a wall!" proclaims David, clinging to his Father's hand. He feels the strength of his Father God. It has been prophesied that by associating with angels, man shall know new music, new architecture, new laws of life.

And the third stone in the foundation of character is chalcedony, further continues John the Revelator. The chalcedony is copper-emerald, strength, by associating with Strength; and sky-tinted opal, emblem of circumambient, quenchless life. Now have I bitten off a leaf from the tree of Immortality. Now have I partaken of Eternity's reviving breath. Now am I wise with high inspiration.

Fortify thy power by the contagion of Power, preached Nahum. He was proclaiming contagion, or for-giveness, beginning at THE SELF, or Jerusalem, seven hundred years before the undestroyable Christ gave orders to declare strength for weakness and life for death; beginning each man with himself.

"Therefore I will look unto the Lord;—Rejoice not against me, O mine enemy: when I fall, I shall arise." This is Micah, stronger than himself by reason of association with Sovereignty.

As the needle is nerved with magnetic power by communing with the magnet, so are we nerved with God-power by converse with the I AM THAT I AM, Author of Omnipotence:

"Thy influential vigor doth reinspire this waiting frame."

Thy lamp of hastening Omniscience shines newly on my liberated brain.

The Hindus drink always the cathartic oil of rejection. The Hebrews quaff deeply of the wines of acceptance. "Deity is best described by negations. Life itself is to be denied till we are independent of life. Substance is to be rejected, till we, like Nanak the guru, are all invisible; sensations are to be disregarded till we can eat live coals, or lie buried in the ground unheeding that we breathe not." This is the religious exhibit of the Hindu.

"Preach the gospel, heal the sick, cast out demoniac dispositions, raise the dead. Welcome the contagion of All-Efficiency. Behold, God exalteth by his power; who teacheth like Him?" This is the religious exhibit of the Hebrew.

"Making the great surrender, Spirit Almighty acts in our behalf," was the Christian discovery of the Spanish mystic Alvarez de Paz.

We can imbue with peace by recognizing the ever-presence of everlasting Peace. We can separate ourselves unto any one force, or energy, or attribute, and by persevering attention to it, can become the embodiment of it. Did not Taglioni become the embodiment of rhythmic motion by practicing it? Did not Margaret of Paris become the embodiment of suffering by separating herself unto suffering? Did not Simon Magus imbue himself with the levitating principle emanating from the earth, by focussing all his attention to it till he was levitated thirty feet in the air?

Now is the time for us to choose the power, or force, or energy, we would embody in ourselves, and make essay at it till all its characteristics are ours, and the efficiency that lies in it is our efficiency. Is it not written, "the works that I do, he shall do also."

"Sing and rejoice, O daughter of Zion! For lo, I come, and will dwell in the midst of thee, saith the Lord." Let us choose to be identified with the Lord strong and mighty, with Him able to keep us from falling whose is the kingdom, the power, and the glory. For our contagion we will separate unto "the great, the mighty God, great in counsel and mighty to work."

The Lord Unknowable All-Knowing, originates all knowing. The Intense, All-Power, is the Author of all activity. The Mystical Stillness wakes tongues. Is it not written, "I will give you a mouth and wisdom, which all your adversaries shall not be able to gainsay"? Was not Stephen charged with Spirit so that he wrought spiritual miracles? Did not Peter speak with resistless eloquence to the converting of three thousand people in one day? Thus have the real mystics of all time been tongues of fire, uncontrovertible, logicians, magazines of scholarship, and the most virile and masculine energies of the age they have invigorated.

Every objective to which we give our attention has

its storage of possibilities ready to spring forth and proceed *in extenso* through its devotee. Tennyson chose
himself as the objective to his inner eye, and made per-
mutations and combinations of the stored vocabulary of
"I, Alfred Tennyson," till he was prince of song, so
strong in his tones that all the world's harsh criticisms
could not drown his supremacy.

Mind follows the visional sense. Hannah's grieved
mind was suddenly transformed to joyous proclamations
of the glory of living, and to praise of the prayer-answering God. But it had taken years of her obedience to the
mandate of the for-giving Presence to have this sudden
transubstantiation by contagion of Divinity; not by
contagion of the levitating principle emanating from
the earth, nor by contagion of the "I, Hannah, elder wife
of Elkanah the Levite."

Jahaziel had elected to identify himself with Miracle-Working Spirit, and at the battle with the Ammonites he proclaimed that the Jews need not try to
defend themselves for his Lord would fight for them.
Like Hannah, he had come to the day of fulfillment.
Schopenhauer chose ascetic morality as the redeemer of
the world. By this choice he finally struck the hard note
by which his name is known: "God is the gigantic evil."
The poet-scholar Leopardi taught mankind that a piti-
less nature has man at its mercy. Through this election
his life, mind, and affairs were chased by poverty, despair
and illness. These are all demonstrations on the blackboard of existence. They prove the transmuting energy
resident in all objectives. They prove that what we now
experience we need experience no longer. Our weakness
waits transmuting for-giveness. Our ignorance stays with
us till our vision toward the Author of Omniscience

comes to us with our own right knowledge. Come, gather to Him ever near, who forgiveth us altogether, so that we find ourselves holding our conversation aright: "O Thou hast for-given myself with Thyself!"

There are unspoken wisdoms awaiting our separation unto the Giver of Wisdom. There is imperturbable health awaiting our separation unto the Imperturbable Author of Health. There is authority over the transactions of daily encounter by separation unto the Unconditioned and Absolute. "No oppressor shall pass through them any more, for now have I seen with mine eyes," declared Zechariah whose eyesight had been baptized from above. "With right glance and with right speech man superintendeth the animate and inanimate." Here we have Hebrew Prophet and Parsee Sun-worshipper heading up with dominance by contagion with the Absolute. Is a man poor, he shall be poor no longer if he but separate himself unto the Owner, Chief Presence in the universe. Is he witless, let him lift his vision, and like Elihu the Buzite, he shall fetch great knowledge from afar. He shall breathe that

> "Breath of heaven all truth-revealing,
> Kindling in him life divine."

We always become negative, soft, plastic, to that objective unto which we oftenest give our inner eye. "God maketh my heart soft," said the devotee to God, Job, the patriarch of Uz. The most powerful of the Babylonian kings, living at the time when metaphysics was the chief study among philosophers, softened his brain by the practice of reasonings such as Parmenides was giving in the science of *Ent.* and *Non Ent.* "But at the end of the days I lifted up mine eyes unto heaven;" he said, "and mine understanding returned unto me—and at the same time my reason returned unto me; and for the glory of my kingdom mine honor and brightness returned unto me; and my counsellors and my lords sought unto me; and I was established in my kingdom."

He does not state how long a time he had spent lifting up his eyes to the King of Kings, the Author of Right Judgment, before he was true exponent of high watch to the counsellors and lords of his realm. But he knew that Daniel had been for three and a half years establishing the Healing Name in Babylon for his sake, and that by finally becoming amenable to its baptism he had identified with Daniel's King. "Now do I extol and honor the King of heaven, all whose ways are judgment; and those that walk in pride He is able to abase." Thus was his proud mind for-given. And thus "Proclaiming Him King, we are happy in His kingdom," promised the Sibyl Cumaean.

A sensitive photographic film steadily exposed to the night skies takes imprint of stars the telescope cannot make visible to the eye of man. So he who becomes sensitive through steadfast attention to the Wisdom-Countenance ever shining upon him knows laws which the books have not recorded. He shows forth activities of unhistoried aspect. "My servant shall deal prudently, he shall be exalted and extolled, and be very high." This is the result of attention to the High and Lofty One inhabiting Eternity, as revealed to the sensitive Isaiah.

To become sufficiently sensitive, negative, tender to an objective, is to be its servant, doing its will. "I came down from heaven not to do mine own will, but the will of Him that sent me." This was the secret of the heroism of the Redeemer. "Ye cannot be negative to two opposites at the same time," was His science.

The triumphing I AM is not vitally promulgated by man, because man has been negative, sensitive, tender to the untriumphing opposite to the I AM. "I said, Behold Me, behold Me, . . . all the day . . . unto a rebellious people . . . that turned their back unto me and not their face." These are the explanations of Isaiah and Jeremiah concerning all who are attentive toward the destructible unlikeness to that One ever offering joy for mourning and victory for defeat.

"Only Thou mayest heal me, Thou most Glorious Manthra Spenta."

In the days when the proud opposites to the Great Fact have rule, shall the King of Kings be chosen—was Isaiah's view of this moment. When *"Labor omnia vincit"* is the motto of men, "Labor not," shall be the risen watchward, "Take no thought" shall be the law. In the days when hospitals are most beloved by reason of their agreement with the destructible opposite, the inhabitants shall stop saying, "I am sick," and the angels shall save all feet from stumbling.

In the days when nations are leaning upon their armies, "Put up the sword" shall be obeyed. While prisons are yawning for criminals, "Neither hath this man sinned nor his parents" shall everywhere be declared true of all men, every eye on the uninjurable Soul Self fathered by Jehovah the Glorious.

While money is the substance most desired, for which kings and scholars are bartering their titles, the sensitive to divine Substance shall "cast their silver in the streets, and their gold shall be removed." While scholarship is at its highest pitch of repute, men shall rise, taught of

God the things that the schools discuss not. "The master and the scholar shall perish." *"In vocavi, et venit in me spiritus sapientia"* shall be each man's scholarship.

Neither shall the powers and capacities of mind be science. "The righteousness of the righteous shall not save him, nor the wickedness of the wicked destroy him," touches a tonic chord of the miracle above thinking and its resultant conduct. God enthroned above the pairs of opposites is the bad man's deliverer and the good man's glorious liberty.

"Thou shalt be hid from the scourge of the (silent and audible) tongue," strikes beyond the range of human effort. "The watchers for iniquity shall be cut off, that make a man an offender for a word," shows a rule of relationship transcending criticism. "Take no thought." "Lift up your eyes," is Christian mysticism. "What I say unto you, I say unto all, Watch," is Jesus Christ magism. "There is a path which no fowl knoweth, and which the vulture's eye hath not seen," saith the Rewarder of the diligently watchful. The "fowl" is the looker for right words and their outcomes. The "vulture" is the searcher for sin and its consequences. This was the patience and the faith of the saints of old, the observers of badness and goodness, the strong believers in rewards and punishments, the unknowing of the law that "that thou seest, man, that too become thou must."

Until the High Redeemer inhabiting Eternity is made the objective of the all-achieving visional sense, he that taketh the sword must perish by the sword, he that leadeth into captivity must be led captive, and no power can ward off the victim's exactness of duplication, for all the time the mystic law is printing on life and mind and body the inner eye's telltale intaglios.

The One served by inward beholding gives for our former nature Its own nature. Milton wrote that converse with angelic spirits etherealizes the body and turns it, by degrees, into Soul's divine essence. Xavier of Navarre, the celebrated missionary, often seemed to be on fire during his prayers to the Supernal Presence. To-day there are those who by contemplating the Healing God rather than their own pains, have had given for their diseased bodies, vigorously healthy ones; for their depressed minds, buoyancy of heart, thus bodily preaching for-giveness.

The thing we fear is the objective our inward beholding touches with contagion. The lightning, the drought, the miasma; loss, deprivation, sickness—they soon find lodgment, embodiment and expression. "Oh! Why will ye die?" cried the Hebrew prophets. Can ye not read the divine decree, "Look unto Me and be ye saved"?

There is a ground of ready affiliation in our constitution, in which the germs of contagion find strong root—God if we contagion God; misery if we contagion misery. Jesus of Nazareth had no consenting ground of affiliation with Satanic cowardice, feebleness, inefficiency. "Evil findeth nothing to tie to in me," He said. His ground He had kept sensitive to the Ruler in the heavens and the earth, so that He could speak forth experiential evidence: "All power is given unto me in heaven and in earth." "Where I am," (O mankind), "there ye may be also."

"Preach for-giveness," He urged. We now know that "we cannot help preaching for-giveness," was the secret salient of the urge. Preach to men to stand and feed in the strength of the Almighty. Preach to them to breathe

in the Almighty; to wake the God-Seed by vision, by
breath; for the inspiration of the Almighty waketh the
understanding, stirreth the God. Agree with One who
is Adversary to pain, misfortune, defeat. Agree with this
Adversary quickly—now!

"Behold I am against thee," saith the Highest Lord
. . . I am against your feasts . . . even your solemn meetings." This is Isaiah feeling horror as of a man dream-
ing, seeing as in a trance the false fodder of low viewing
with which his beloved companions are feeding to their
destruction. He senses the now well-known law that we
feed on what we inwardly behold. He sees the saving
Substance that his neighbors might feast upon by lifting
up their eyes to the fields white for the harvest, but he
cannot make them look and taste; like as in a Sibylline
dream, he speaks for the waiting Substance. His neighbors know his vision is Truth, but they obey not. They
were uninformed that long before Isaiah's time, the Parsees had proclaimed that nine hundred ninety-nine
thousand, nine hundred ninety-nine diseases spring from
low visioning; but that the most Glorious Highest being
sought, the diseases should all fall away.

"The ransomed return with singing." They reach
joy, affiliating with the Author of joy. The taste of joy
gives sense of success with a new leverage. The deaf man
who spoke with passionate earnestness to the Presence
of God in the universe, felt a great cleavage in his head,
and blood flowed forth from his ears. The gates of his
imprisoned hearing being opened, he shouted for joy.
The kingdom of heaven cometh and findeth something
to tie to in him who touches the God-estimates *Scire*
meets *scire*—knowing meets knowing.

Honor and fortune and knowledge are for-giveness

by recognition of the Glorious Presence of Victorious Divinity. Health and strength and joy and peace co-operate as God with man, by upward visioning. Did not Dionysius the friend of Paul find new knowing by upward viewing toward the Author of knowing? New knowing starts with mystically sighting toward the Original Knower by whom our knowing roots are quickened. Let us know from our own base, and poverty and foolishness and evil have nothing to tie to. Thus do we find our original goods. So are we for-given. For, "Behold, God exalteth by his power; who teacheth like him?"

Paul called the day of joy the day of atonement. "We joy in God through Jesus Christ, by whom (accepting our sonship, as He declared) we have now received the atonement." Jesus called it the day of for-giveness. To the man let down through the roof, sick with the palsy, He said, "For-given." To the woman bruised of heart, He said, "For-given." And thus were these both brought to conviction, not of sin, but of Sonship.

Agreement is harmony. And harmony with the Adversary to pain, ignorance, disorder, means success like the Adversary. "I will contend with them that contend with thee. He that striketh at thee striketh at Me. Thou art My servant, fear thou not. They that war against thee shall be as naught." "O, I Thou, and Thou, I!"

High success denotes entire harmony, entire forgiveness. He that is entirely for-given speaks with resistless inspiration. He has the *hestia vestia*, the heavenly hearthfire. He is a world kindler. No opposition daunts him. Like the deformed French child who knew that the doctors in the hospital would cure her, in spite of their "knowing they could not," till the unbelieving doctors did indeed cure her, so we, for-given, know that life,

health and joy are eternally native to us all, and our
positive fervours warm past all doubts. The third angel's
burning Lamp is the speech of the Imperturbable
Knower. It is mind tonic, wormwood to the vitals, ware-
mood, mind-preserving acquaintance with the presence
of the Healing Christ.

"He that humbleth himself shall be exalted." He that
letteth himself go to the Finished Fact, as the inconsequent needle yields to the magnet's empowering, is a
new character on the earth. By his utter meekness he
is liberated from himself, and works the works of the
Worker unto whom he has yielded himself.

The Sacred books, uttering the inspirations of the
God-taught, the *Theophoroi,* lay great stress on voluntary
surrender to the Divine Trend. "Put on humbleness . . .
meekness." "Because thine heart was tender, and thou
didst humble thyself before God . . . I will for-give . . .
and heal all their land." "Thou shalt walk prosperously
because of meekness." It was while Daniel was voluntarily
casting himself down, to be taken possession of by the
Saving Sovereignty in the Universe, that the angel
touched him with heavenly inspiration and said, "O,
Daniel! I am come to give thee skill and understanding.
Stand thou upright on thy feet."

We gladly offer the sum total of our unlikeness
to the Almighty Giver. We gladly offer the initiation fee
of our contrary tempers at the courts of the Healing God.

The third angel's voice wakes the will to let go the
last vestige of opposition to the Mighty Trend. What
matter how unlike to *our* way is *The Way* our life seems
to be moving? We take with us words and return, looking
steadfastly unto the Great Mover. Is it not assured that

when the Lord returns our returning, our mouth is
filled with laughter, and our tongue with singing? How
else than by being free inspiration can we warm the world
into health? How else than by being God-glowing can
we go into all the world preaching the gospel and raising
the dead? How else than by High Association can we
contagion free inspiration, the Holy Ghost influence
that sweeps down all aftermaths of low visioning?

The Hindu sometimes touches the law hymned
by the third angel:

"Bow down to Me, and thou shalt come even to Me
. . . Take sanctuary with Me alone . . . I shall liberate
thee from all sins by the resplendent Lamp of Wisdom."

Milton immortalized his acquaintance with the third
star's supernal import:

> "What is dark in me
> Illumine. What is low raise and support,
> That to the height of the great argument
> I may assert Eternal providence,
> And justify the ways of God to man."

The all-conquering Jesus passed through the gates of
voluntary lowliness, and taught us all that identifying
worship, when it touches the conquering truth, has come
up out of the baptismal font of humility. "God *is* a
Spirit: and they that worship him must worship *him* in
spirit, (of humility) and in (the bold words of) truth."

Now we are ready to cast ourselves and all our wills
and demandings in lowly yielding up to Unseen High
Sovereignty and His own Providence. Who is not glad
to surrender his proud mind's muddy wadies of foolishness, its dark pools of ignorance?—

Here is my mind, I spread it out before Thee. For-give Thou its foolishness and ignorance with Thy bright wisdom.

Here is my life impulsion, I offer it to Thee. For-give Thou all its contrariness to Thee.

Here is my heart; it is Thine only. For-give Thou its dissatisfactions; for-give its restlessness. For-give its dis-couragements; for-give its elations. For-give its hopes and its fears; its loves and its hates.

Here is my body, I cast it down before Thee. For-give Thou its imperfections with Thy perfection.

For-give me altogether with Thyself. So only can I be the life and inspiration of the five bold words of Truth: Hymns to the Eternal—glowing Virgins with oil of healing and oil of illuminating in their everlasting lamps—

1 Thou art and there is none beside Thee, in Thine own Omnipresence, Omnipotence, Omniscience.

2 I am Thine only and in Thee I live, move and have being.

3 I am Thine own Substance, Power and Light, and I shed abroad wisdom, strength, holiness from Thee.

4 Thou art now working through me to will and to do that which ought to be done by me.

5 I am for-given and governed by Thee alone, and I cannot sin, I cannot suffer for sin, nor fear sin, sickness or death.

My soul, doing obeisance unto the Wonder of Thee, wakes again these hymns of the Morning Stars in praise of Thee.

High praise of Him: all For-giveness draws hitherward the promised New Language; and brings into view
"The New Race to be sent down from heaven" foreseen
by enrapt Sibyls, keepers of the five hymns that should
some time "sing-in" The Golden Age.

This Fourth Study has treatment quality for all who read it, even though they may not have heard its subject matter discussed orally. Practicing its lordly formulas wakes victorious energies.

E. C. H.

IV

FAITH

Every number held profound significance to the ancients. Number *four* held the fire of convincing energy. It was the Uriel Angel of divine telepathy. Beresford, the English writer, declared, that he caught belief in survival after death from the mass faith at a meeting of Spiritualists. He did not report that the sparkling up of the faith center in man is the waking of his hidden miracle-working genius; the great outfiguring of *number four* to Pythagoras, sign of the fertile square according to the Cabala, the touch of fourth dimensional strength, the change from Moses meek to Osarsiph bold according to Egyptian Hermetics.

The fourth stone symbolic of character according to St. John of the Revelation is emerald. It was once called smaragdos and held radiations for sharpening the memory, even to the recalling of our heavenly beginnings, making us mindful of "that country whence we came out," as Paul wrote to the Hebrews, assuring them of it as a country to which we all may return.

> "Not in entire forgetfulness,
> But trailing clouds of glory do we come
> From God who is our Home."

Everything about *four* was fourth dimensional to the wise men of old. Notice them telling how man is comraded by angels from the city of God when he finds himself touching the fourth side of the city that lieth foursquare. *Things* have never satisfied his seat of sacred starvation; nor yet noble *thoughts, high statements,* even the highest; nor practice of ectoplasm and its astral shadows of departed friends. Only by laying hold of the High Adequate has man laid hold of that which satisfies his heart's desire. Notice the wise men telling of Jacob by the Jabbok brook sensing the angel of God who called himself God, changing him from Jacob the frightened to Israel the fearless, and causing him to found a dynasty of kings ending in earth's final King—*The Nazarene Jesus!*

The phoenix bird which fell into helpless ashes and rose into winged majesty was once the symbol of man's helplessness in the face of death, rising into daring renewal above death by the sacred touch of heaven's Uriel fire on his yearning heart's despair. Not only did the phoenix signify survival after death but revival out of death, even as the King of Judah rose while yet Isaiah the mighty was laying the ban of death upon him.

As the mariner on the sea steers his ship's course by a needle which points to a magnetic north, not to the north of polar bears and icebergs, so man is truly steering his hopes by an inner needle pointing to a country unseen from whence in time of danger or despair miraculous succor may swing toward him.

Something within us innately hopes great things from the self-existent kingdom to which King David turned crying, "Mine eyes are ever toward the Lord, for he shall pluck my feet out of the net;" to which the sage of India gazes and is touched with long life because the kingdom is ageless.

David was rewarded for his bold insistence, his persistent high watch—"We went through fire and through water; but thou broughtest us out into a wealthy place," he cried.

It was the business of the Levitical singers in David's time and in Solomon's to sing the ways of the kingdom unseen in its miraculous workings with this visible world and its people. "Thou shalt ride prosperously because of meekness," they chanted to the high-pitched, rich-toned sackbut of many strings.

David had been meek even to sorrowing daily in his heart before the Lord of his hope who seemed sometimes to hide His face from him. Therefore was the promise fulfilled upon him, "Thou shalt ride prosperously because of meekness." "I will sing unto the Lord because he hath dealt bountifully with me," he triumphantly proclaimed.

There is a mystery about meekness, gentle receptivity, which even the merchants of Rome and Athens knew, as before the Christian era they bowed their heads before unseen Mercury the god of magistral to poverty. And as Hesiod the Greek taught, bowing his head to angels that they might sprinkle him with wisdom-glory or with gold, not according to his human will but according to their own heavenly decretals.

Sometimes we read that those astral pictures which Homer called shades, may meekly be yielded unto, but truly if there is presence of King of Kings and Lord of Lords with givings and workings supernal ready to let fall upon our human lot, why choose shades of the dead? "Hast thou faith? have it to thyself before God," had better be our starry choice on this our plain of Esdraelon.

So shall He give His angels mysterious ministerings in our behalf!

The Levitical singers of David's time and of Solomon's time sang that the daughter of Zion should be Uriel-fired with kingship. "Daughter" was the Levitical singers' figurative word for the Most Meek among the people. Was not the daughter of the old Hebrew house the most meek member of the family? Was she not handed over to her husband as docile and adoring, as seeing in him "her lord, her governor, her friend?" Was it not recorded that her confidence in his greatness caused him to be known in the gates, when he did sit among the elders of the land? As lighted candle lights candle, so conviction fires conviction. Elisha was lowly in his conviction of Elijah's Godlike greatness as head of the schools of the prophets of Jehovah in Gilgal and Jericho. So Elijah, showing forth that Godly majesty, touched Elisha's meekly receptive being with conviction of competence, and he rose up head of the schools of the prophets of Jehovah in Gilgal and Jericho.

Long before the time of Elijah and Elisha it had been taught in mystic language that we rise up with that authority before which we have been meek. Was not Isaiah meek before the Lord of hosts till the Lord of hosts told him to command the Lord of hosts as an Obedient Servitor? Was not Jeremiah meek before the Ruler in the heavens and the earth till the Ruler in the heavens and the earth told him to show himself ruler over the nations and over the kingdoms?

Did not Jesus mean, "Have the Rulership of God Himself" when He told His disciples to have the faith of God? For is not *faith* rulership? Is not *faith* kingship, or confidence to command? Is not kingship always associated with confidence to command? "If ye have faith as a grain of mustard seed, ye shall say unto this mountain, Remove hence to yonder place; and it shall remove; and nothing shall be impossible unto you."

Peter found by his own obedience to the bright angel who smote him on the side, saying, "Arise up quickly, gird thyself and bind on thy sandals . . . cast thy garments about thee and follow me," that the bright angel was obedient to him, opening the barred gates and loosing the chains of the four quaternions of soldiers to whom he was bound. It is no wonder that Peter wrote it down for an eternal verity that angels, authorities and powers are subject to the hidden man of the heart; the waiting authority principle lingering in the being of every man, woman, child, on earth.

The mystery of obedience to authority as surely rising as authority is every instant manifest. Do we not have to obey the authority of the door knob before it works for us? Or have to obey the rigid law of our feet before they do what we wish of them? So the Mighty King we call God gives orders to which we must yield obedience before His sublime service in our behalf is sublimely manifest: "The Lord lifteth up the meek"—the gently receptive to burning God conviction, which is confidence to command, which is kingship ever waiting to find its meek sparkling tinder within us.

Was it not wonderful of the gentle Japanese to discover that if one had faith, which is confidence to command, no larger than the point of a needle, he could say to a dead sardine's head, "Walk me over the water," and it would obey? Was it not astonishing that Count Puysegur of Buzancy could rouse up confidence to command a strong tree to heal all who touched it, and it meekly did his bidding? Was it not mystic inspiration in Maxwell the Scotch metaphysician to find that he could rouse confidence to compel the secret Spirit of the universe to do blessed healing ministries for him?

Maxwell did not know that he was practicing inborn, native kingship by such bold commandings; neither did the Japanese, nor yet the hundreds of daring new missioners, who go about the world saying to the lame, "Walk!" and they walk. Or saying to the deaf, "Hear!" and they hear. But none of them can tell us like the Hebrew prophets and the masterful Nazarene the practice that rouses the living dominance called by Peter the hidden man of the heart, our secret Jehovah Nissi (Jehova my banner).

The prophets and Jesus teach us that being insistent and firm with the Waiting Adequate we shall find the Waiting Adequate most willing and competent. "Is anything too hard for me?" He saith. "Hast thou faith? have it to thyself before God." "Lo, I am with you alway." "For the Lord shall be thy confidence, and shall keep thy foot from being taken." "Concerning the work of my hands command ye me."

Note how universal God majesty awaits the rise of man majesty universal!

Joseph in the prison house of Pharaoh of Egypt was meek to the fulfillment of the prophecy that he should save the Jews from starvation. He stopped his own thinking for the Unseen Knower to strike the hour for divine wisdom to touch his brain with words not known on all the earth. So great Pharaoh set him over all the provinces of the realm and gave him the handling of all the gold and silver of the realm, and today every Jew on earth owes his life to meek Joseph rising to kingly authority by reason of being touched with sprinklings of gray matter from above till his speech did distill as the dew.

Napoleon was also an example of letting his own thoughts stop for the thoughts of those higher in authority to sift on his brain. Catching their dominance he proudly said, "The only difference between me and other men is that I have confidence to command." It was not till he began to study the science of battles that he lost victorious confidence caught from Victorious Confidence. We find on looking over the people on this earth who have been baptized with originality that they have let the world's thinkings alone, and even for no telling how long have stopped, perhaps unwittingly, their own thinking also, and so creative new knowledges have been free to touch them. Here we come upon the magic wisdom of Jesus of Nazareth: "Take no thought"—"In such an hour as ye think not."

We even read how some clergymen admit catching their thoughts from the thoughts of their congregations. So they are not original in their instructions. The world now needs fresh news from Universal Wisdom. Who can stop studying Latin enclitics and ages-old vivisections long enough to bare his meekness to new distillations from Divine Beneficence, sparkling gray matter-drops charged with healings from on high? Has any heavenly distilling reached mankind from the Sultan Amurath's striking off one hundred Persian heads that his physician Vesalius might watch the spasms in the muscles of the human neck? But note what the voice of inspiration declares to Amurath and Vesalius: He that taketh the sword must perish by the sword. He that leadeth into captivity must be led captive.

High faith is confidence to command the Working

Executive, standing up in the universe to the point of hearing as good response as Jacob, forebear of royal Jesus, heard: "As a prince hast thou power with God and with men, and hast prevailed."

Or as John of the Christian Apocalypse heard, "And there shall be no more curse."—Karma, consequence of past actions or thoughts.

Jesus discussed the mystery of forgiveness. He proved the mystery of bold use of the Working Executive facing us through all things, ever saying, as Iamblichus discovered, "Boldly tell Me what to do and when to act."

When shall the fig tree, symbol of all flourishings, fruit for the one who discovers his own bold authority? Never! When shall the fever desist for such an one? Now!

To what was Joshua speaking when he stopped the sun and the moon in the midst of the heavens? To the Lord facing him, as we read in the book of Joshua, tenth chapter. To what are the little children of India speaking when the sticks and the stones with which they are playing do actually move here and there at their orders? To the same Lord facing them that faced Joshua the daring "I am captain!"

Why did not Bjerregaard go on with his discovery that "the earth is creating and destroying *because it knows no better,*" and boldly tell it better, as the Jewish Bible with its vigorous miracles offered him precedents?

"Prosper thou me!" commanded King David. "Prosperity is of thee." "The silver and gold are thine." "Riches and honor come of thee."

Such truthful recognitions caused plenteousness of gold and silver to come to him exactly as such truthful recognitions would now cause plenteousness to come to any one of earth's multitudinous sons of the Highest.

By this fourth lesson with its grand offerings, we see that
Deity is no disciplinarian giving us hardships and refine-
ments of deprivings, but a Beneficent Presence awaiting
our use of everywhere-facing-Beneficence by bold insistence, like the "Glorify thou me" of Jesus; the "Prosper thou me" of David; the "Answer thou me" of Job;
the "Stand thou still" of Joshua.

"Come boldly up," said Paul. Why not come boldly
up if "boldness hath genius, power, and magic in it?"

This One everywhere and through everything facing
us is no "hound of heaven" hounding us to starvation,
cold and death. Neither are we His hound dogs beaten
into submission to His ceaseless disciplines. Let us take
right view of Him: "Ask what ye will," He saith. "What
wilt thou?" He asketh. "Concerning the work of my
hands, command ye me," He urges. "Is anything too
hard for me?" "I will work, and none shall hinder."

When the Belgian writer tells us to be frank with the
God Presence and tell Him we are dissatisfied with our
lot, the Belgian writer does not seem to know such assurances multiply our dissatisfactions because they pick
up the formulating substance charging the ethers and
embody according to their recognitions. Tell him to
speak boldly, looking into the face of the answering Substance, "Deliver Thou me from evil!" "Give me this day
my super-substantial bread!" "Give me courage, confidence to insist! Bless me with life, wisdom, divine efficiency!"

Tell him this recognition picks up the formulating
substance and translates it into the mystic's fulfilled assurance, "So shall thy life renew; so shall inspiration teach
thee; so shall thy affairs go newly right with thee." We
light our inner vision by exalting it. Lightened vision

wakes all our faculties to sense the Supernal Good-Will-
ing surrounding us, forever wooing our positive, "Give
me for my weakness, strength to command Thee!"

Some things will never square right with man till he
takes Deity at His word, "Command ye me."

Stop talking *about* God and His idea man, and speak
unto majestic Deity face to face! So shall majestic man
arise, victoriously daring!

—b—

As Adam and Eve were not only individuals but
perceptions, so are the angels of the Apocalypse not only
winged messengers but high perceptions and their activities.

The Egyptian Magi changed the name of the neo-
phyte at the fourth perception, because at this his nature
changed. From being a meek listener he became a bold
speaker; from being a timid follower he became a daring
leader.

The fourth angel smites one side of the sun and on
that side it is dark. So did the same angel smite Jacob and
one side of him was withered, not for use but for super
use. So did this mighty angel smite Peter in the prison,
and the smitten side of him being now supernal perception and not common intelligence, even as angels
of the Free Adequate, opened the bolted prison doors
and undid the chains and manacles that human animosity
had welded. The Roman soldiers guarding him, four
quaternions strong were smitten, and the miracle pro-
ceeded onward uninterrupted.

The fourth perception, setting aside the common law,

exposes the unmanageable fourth dimension in space, which makes locks and bars and lions' teeth and adverse criticisms of no account.

Job the stricken was searching for help with his watch toward heaven, when suddenly he sensed the fourth dimension, and life for him became a track of victorious light to lighten all generations after him.

Jacob sensed the presence of the angel of the miracle, the angel of the helping, and wrestled with the angel, enduring as seeing the invisible, and his name was changed to Israel. He was no longer Jacob the cringeling, but Israel the Prince whom God Himself served. To any daring wrestler with the ever present angel of the miracle any man may hear that Angel Servant responding, "Concerning the work of my hands, command ye me." "As a prince thou hast power with God."

At the fourth perception David found the same Servant: "Bow down thine ear to me; deliver me speedily," he cried. "Thy gentleness hath made me great," was his astonished acknowledgment.

At his fourth perception Isaiah implores all mankind to practice the formula of the fourth dimension, whether they themselves have been entranced by the fourth angel's smiting or not. "Thus saith the Lord, Ask me of things to come, and concerning the work of my hands, command ye me."

Jesus the Redeemer gave the formula of the fourth verbatim. It is the speech of the fundamental knower risen up out of the waters of humility. It is the speech of the hidden man of the heart, without age and of no nationality. It is the genius of Massini at seventy singing Gounod's "Sanctus" to an enthralled congregation. It is the genius of Elman at seventeen drawing a magic bow across a magic instrument to enraptured throngs. How thirstily the people put their lips to the troughs where living waters flow! What hearts of love they lean close to fires celestial!

All the world travails for the fourth angel's birth-mark—the parting of its common mind for its uncircumscribed genius to act. "Who is this that cometh from Edom with dyed garments from Bozrah? . . . I that speak in righteousness, mighty to save." I that have dyed my language in the word of the High Supernal. I that have dipped my will in the Heavenly Trend. I, smitten by the angel of the miracle and his delivering might. "Arise up, quickly!" the angel says. Now am I as Jacob, not for visible but mystical usefulness. Now am I as Peter, free Spirit.

A principle is a comprehensive law or doctrine from which others are derived. That is, obedience is vested in the Supreme I AM or there could be no obedience in the dog or horse. Authority is resident in the King of Kings or the General-in-Chief of an army could not command with success.

When Iamblichus of Chalcis found that the weather obeyed him, and eagles flew hither and yon at his insistence, he supposed there must be an order of obedient invisible powers in the universe, altogether at the bidding of man.

By reason of triumphs which certain men of old achieved after speaking with commanding determination to their invisible gods, they sang:

> "Cease your fretful prayers,
> Your whinings, and your tame petitions;
> The gods love courage armed with confidence,
> And prayers fit to pull them down. Weak tears
> They sit and smile at."

Something concerning the mystery of man's inborn authority has ever been the fourth theme of such as have consciously or unwittingly obeyed the Supreme Edict, "Look unto Me." By snatches of what Luke the Apostle called "sunrisings from on high," the illuminati of the ages have known that the will to command the Obedient Supreme Presence rises up after obedience to the will of the Supreme Presence.

"If man avoids regarding himself as king of the universe it is because he lacks courage to recover his titles thereto," wrote one of the illuminati after having been by meekness dissolved into recognition of the Majesty of the Commanding Supreme, and felt its quickening stir as likeness triumphant in his own breast.

The law is plain enough. If that nature before which we have been negative, soft, meek, plastic, draws forth and stirs alive in us its own kind, it is not surprising that the meekest and lowliest of all men rose up with the bold proclamation: "All power is given unto me"—"I have overcome the world." It is not surprising that His disciples, catching His assurance, found that satanic tempers fled at the sound of their bold commands, and the willing angel of the miracle stood by them to save them from prisons and swords.

"Tell ye the daughter of Zion, Behold, kingship cometh in meekness." This is Zechariah agreeing with Jesus across the gulf of centuries. The mystic law is one and its way is one as mathematics is one. Does the relation of the hypothenuse to its base and perpendicular ever alter? Pythagoras sacrificed an hundred oxen of rejoicing when he discovered that eternal relationship of the hypothenuse. Jesus gladly sacrificed Himself to call attention of mankind to the root of Divinity, the spark of identical substance with the Unconditioned Absolute inherent in them each and all.

Job's saying that the root of the matter was in him had not sufficed to call the attention of men to their own Absoluteness. The prophet's assurance that "He hath made of one blood (or root and stalk) all the nations," had not given the serf and bond woman inkling enough of their right to dominion over that mysterious Servant-hood standing up in the universe.

As in mathematics the time came in with Pythagoras for knowing that the root of the sum of the squares of base and perpendicular was forever the diagonal, so in with Jesus came the time for showing the root of divine authority bone of bone in men forever, in their relation to the Supreme Good Will occupying Omnipresence: "Thus saith the Holy One of Israel, and his Maker . . . concerning the work of my hands, command ye me." (Isaiah 45) Therefore, after this manner pray ye:

Give me this day my super-substantial bread—bread for my eternally innate authority with the God that standeth in the congregation of the universe!

When Saint-Martin tells us that it is lack of courage that keeps us from acting with kingship, he does not tell us how to rouse that courage. When Jeremiah was shown that it was a sign of arrested development to tarry as a cringeling in the face of the waiting Good Will, he did not understand that he was to instruct all the Jews in rousing their courage to speak as lords of the Obedient God. He heard it as for himself only, "Say not, I am a child—See, I have set thee over the nations and over the kingdoms—to throw down, to build, and to plant."

But Jesus, the Bloom in the Garden of Man, rising up out of authority-breeding lowliness, said, Speak like masters to the responsive stately God of Lazarus; to the stately responsive God of the man with the impotent arm; to the stately responsive God of the mountain; to the obedient responsive of the sycamine tree. After this manner speak ye: "Thy kingdom come! Thine is the Kingdom forever! Stretch forth thine hand! Make straight the path!" His God did not over-discipline man. His God awaited man's bold insistence, "Make straight my path!"

This is the rise of the Hidden Man of Job, of Joshua, of Jacob—the great triumvirate of J's on the commanding heights of courage to command the Willing Omnipotence ever whispering to all mankind, "Concerning the work of my hands, command ye me!"

Did not Job hear the Supreme Authority in heaven and earth speaking with sternness, "I will demand of thee and answer thou Me"—over and over, till the intone of it smote his root of divinity, and he turned with the same address, "I will demand of Thee, and answer Thou me!" And is it not recorded that the Lord was pleased with Job?

Was it not to the Lord fronting him through the sun and moon that Joshua spoke with bold commanding, "Sun, stand thou still upon Gibeon, and thou Moon, in the valley of Ajalon! . . . So the sun stood still in the midst of heaven, and hasted not to go down about a whole day. And there was no day like that before it or after it . . ." when Joshua spake unto the Lord. With the rise of his inborn root of authority spake he to the Willing Obedience facing him as the Omnipotent One!

And did not Jacob wrestle to give his hidden boldness dominion? "I will not let thee go except thou bless me!" And the yielding Angel of Victory did vouchsafe the

blessing. "I have seen God face to face," said the transmuted Jacob. I was afraid, but fear had no annulling
strength against my vision of God.

Napoleon Bonaparte was docile and promptly obedient to his superior officers; watchful of their genius at
commanding, till his latent generalship stirred and he
turned on them all, head of the army of France.

Hannibal was from infancy meek and plastic before
Hamilcar, General-in-Chief of the Carthagenian army.
Every hate and every love of Hamilcar stood forth in
Hannibal at its proper moment. Is it surprising that at
twenty-eight he is head of the Carthagenian army like
his father?

How docile was Joan of Arc to the wills of the angels
with whom she had converse all her life, dauntlessly repeating their directions to the awe-struck generals, soldiers and statesmen of France, till she at the age of six-
teen was *tête d'armée*.

We must choose well the objective before which our
inner eye oftenest pauses, for if the objective has not commanding boldness, resistless authority as its savor, when
the moment of identification transpires neither will there
then be any commanding boldness, resistless authority
rising up out of its sleeping place in us.

Gather a hint from the slow rising Moses, docile,
teachable, tractable, before the tutors of princes in
Heliopolis at a time when the tutors and priests of Heliopolis were famed for their learning and manners. Is it not
written in secular history that Moses also was famed for
his learning and manners? But at forty years of age he
fled like a cringing Jacob at the threats of two Israelites.
No confidence to command and be obeyed had leaped

like a fountain of fire from its slumbering pit in him.
How could it, if the tutors to whom he had been religi-
ously attentive had never waked their own fearless dominion? Can a stream rise higher than its source?

Now, as an exile among the mountains of Midian, he
has spent forty years humbling himself before the High
Deliverer, the Noble Counsellor, the Almighty Champion, and though he is eighty years of age two million
Israelites obey his lightest word of command. Notice the
mathematical increase, afraid of two, dominant over two
million! The Lord of Lords and Ruler in the Heavens
and on the earth sends him forth Lawgiver, Governor,
Mighty Champion, High Deliverer like Himself. He
sends him forth with youth in his genius, the stamp of
Fadelessness on his body.

> "Here eyes do regard you
> In Eternity's stillness:
> Choose well, your choice is
> Brief and yet endless."

"Let the Lord be thy confidence, he will not suffer
thy feet to be taken." This is the principle of attention
to the Highest Lord to the point of rising above prisons
and lions' jaws. This is the principle of making them of
no effect. The lordship that causes the iron gates to open
of their own accord, that rolls away the stones from the
pathway, must hail from above the three dimensions.

Joan of Arc was left to be burnt at the stake at twenty
years of age, because her vision had not sought higher
than the faces of the flying messengers. Napoleon is in
common exile at forty-seven, because he has never sent
his vision higher than Emperors' faces and heads of

armies for its snatches of quickening dominion. Hannibal is in durance at sixty for the same reason. But Jacob and Elisha and Paul finish their course with the words of light still on their lips, and the crown of the conqueror still shining on their heads.

And the kings that shall arise after them shall be lords even over the Sabbath, or the stopping place of death. "There is no Sabbath keeping in the temple," whispered the rabbis. The Lord of the temple is Lord. He says to the obedient Executive standing still and tall in the flourishing fig tree of fever or dying, "It is finished!" and nothing can resist the Lord's command to the eternally present Obedient God.

The rise of boldness, authority, is the rise of inborn superiority to surrounding conditions. It is wresting the tongue from outward descriptions to conform to heavenly fact. Authoritative speech brushes aside the cobwebs of outward appearance. It is backed by the mystery of the conquering kingdom of the Inmost Actual.

When Solomon said, "The opening of my lips shall be right things," he meant that he would speak forth from the hidden man, as free Spirit that knows nothing of defeat or poverty or sickness. By this speech he would lift his head above conditions. At a certain moment the hidden man of the heart, gifted with dominion, leaps like lightning to expose its magical independence of the length, breadth and thickness of matter, mind, sensations and their world maneuvers.

Does Habakkuk say that he yields to grief when the fig trees blossom no longer in Judah? "Although the fig trees shall not give blossom—the labor of the olive shall fail—yet will I speak rejoicings—my feet shall walk on high places—let the chief singers chant with me!"

It is the failure to stand by the things of Almighty Spirit, to the leap of authority, that accounts for the seemingly unmanageable misfortunes of aspiring men in all ages. They have supposed, in unguarded moments, that the yielding they must make was to the overbearing three dimensionals of misfortune, old age and dying.

Let us heed the voice of inspiration. That yielding which the sons of earth are dimly tending to make, is not to the three dimensions but to the God law that works above them. Although now, apparently, by my past downward viewing I have walled myself into feebleness, sickness, defeat, yet, speaking boldly from my bright secret Self, I am Strength itself—I am flawless Confidence, I am Master of the willing God of my universe. This is the "opening of the lips with right things," and all the divine forces stand ready to minister to my leaping Word.

As the young eagle presses his leathery joints against the cracking shell, all nature waiting in mute sympathy, expecting to be governed by his new born demands, so the Still God of the universe waits to move through all visible and invisible items to minister in willing docility to my undiverted high confidence.

Although "the flocks shall be cut off from the fold, and there shall be no herd in the stalls; yet will I joy in the God of my salvation," still forever we hear the voice of the poverty-surrounded Habakkuk singing through the night watches, our steadfast example through the ages.

The meekness of the mind, the will, the heart, opening to the Healing Good, is moving aside for a lordship not of the flesh to act. "Watch therefore; for ye know not what hour" your lordship rises.

High Mysticism

"As when by drastic lift
Of pent volcanic fires,
The dripping form of a new
Island springs to meet the airs,
So from our deeps we rise."

"Now will I rise, saith the Lord; . . . now will I lift up myself." "At the lifting up of thyself the nations were scattered."

It is the rise of the Divine Will to see, when the blind beggar throws aside his ragged garments and runs to the waiting Jesus. It is the rise of its pent up fragrance when the tightly closed petals of the rose fall back and the hidden splendors of color and perfume face the sun, uncramped forever.

"Dost thou ask what Christianity is?" says the Mohammedan Sufi, forgetful of creed and country: "I shall tell it thee: It digs up thine own ego, and carries it up to God."

It is the rise of the divine ego that makes a man victoriously bold. "Come boldly unto the throne," said Paul. "Boldness hath genius, power, and magic in it; what you can do or dream you can, begin it. Therefore be bold!" Persist like Jacob.

Though my low views have sent me loss of friends, pain, humiliation, yet truly am I strong son of God, with dominion in all my vital sap. I am at my roots greater than my environments and the shadows of hardship with which by turning from the Highest I have darkened my path! And Omnipotence stands before me and behind me, at my right and at my left, above me and below me, to serve my rising commandings, as He Himself hath voiced by priests and prophets, and the young eagle's springing.

So we are to look upon the Man who threw aside the wrappings of the grave, the stone-sealed tomb and the soldiers' swords, bursting their three dimensional bindings with risen divinity, as law for the whole of us, world without end.

After this commanding manner speak ye: Let me not turn aside from facing Thee! Deliver Thou me from evil. Thou art empowering Obedience. I owe Thee bold command, O Thou Owner of all the kingdoms!

There is a noble triumvirate of D's on the self-authorizing rock of conquering confidence: David, Daniel, Darius—"Show me a token for good, that they which hate me may see it"—"Let my Lord speak; for thou hast strengthened me." "Thy God whom thou servest, continually, he will deliver thee." They show how at the first upspringing of this confidence, this bold certainty, the God in the universe serves promptly. The symbol of this upspring is the emerald stone; stone significant of walking free from common law, unified with the miraculous, where he that would hinder thee cannot discover thee.

This is the science of high visioning—of looking unto the Vast Vast Countenance with healing of our tardy recognition of our own inborn kingship as its fourth gift.

Is not faith the gift of God, according to Scriptural instruction? Is not faith the confidence of things chosen according to the same high information? And does not masterfulness rise with confidence? And are we not told to have the faith which is the masterfulness of God himself? "Have the faith of God," said Jesus.

"Thou hast a strength of empire fixed."

The exaltation of lifting up of the vision is "Fear of the Lord." "Pass the time of your sojourning here in fear," preached Peter. It is written that the fear of the Lord is the instruction in wisdom. It is written that it is the beginning of wisdom, or light. If thine eye seek the Lord only, thy whole body shall be full of wisdom. If thine eye seek the Lord only, He will fulfill thy desire. If thine eye seek the Lord only, He will be thy strong confidence. If thine eye choose the High Deliverer, thy dominion shall rise up.

Thus have the inspired among men written in their own risen moments, always showing by their instructions that their risen kingship stirred forth from the bed of lowly-heartedness. And always lowly-heartedness before Supreme Majesty, else how should kingship rise with its sceptre? "To this man will I look, even to him that . . . trembleth at my word," saith the Supreme Lord.

Jeremiah trembled at sight of the danger streak in the Jews' sum total of character. The wicked and foolish trait deplored by all the Jewish wise men had glued Jeremiah's unmitigated attention. Out of the molten depths of his anguish he forged the prophecy of doom in which that trait would lawfully ultimate. It has taken centuries on centuries for the Jews to labour out from under the black bar of Jeremiah's decree:

> "Your inheritance shall be turned to strangers,
> God is wroth against you, O people of Zion!"

Jeremiah sometimes forgot the sin streak of the Jews, and looking above them saw them above themselves, for seconds of time, as the High God saw them:

"In those days shall Judah be saved . . . for thus saith the Lord; David shall never want a man to sit upon the throne . . ."

Note that Jeremiah prophesied coming greatness and glory while his vision was toward the High and Lofty One that inhabiteth Eternity, who cannot Himself look upon evil. Seeing sometimes as God sees he sensed the liberation of the Jews from the stream of their forefathers' sins, but never long enough to sense their right to their present Victorious Sonship to their Heavenly Father.

Jesus the star out of Jacob, bright with the morning of the liberation, told them that no man upon the earth was their father; one Only was forever the Father of all, even God. He dipped his speech in the truth of high birth and victorious life. "Neither hath this man sinned nor his parents"—"The flesh profiteth nothing"—"And they shall see His face—And there shall be no night"— "Go ye, and make disciples of all nations."

He taught that he that is steadfast unto the day of believing, commanding confidence, faith, shall be saved from the law of cause and effect, the karma of past vision and thought. For though foxes have holes of resting, and birds of the air have places for their pause, the risen soul belongeth not to their order. There is no set outcome to the vision toward the Infinite. Though "envy is rottenness of the bones," the loss of envy has no stopping place of freedom. Before the Son of Buoyancy the doors are all open. Though "the hypocrite's hope shall perish," forever, in some particular disappointment, the loss of hypocrisy has no limit to its good and perfect gifts from above.

The great aphorisms of men wrought out from looking away from the Highest, cease from being true, but the high truths that belong to the Vision above the aphorisms of downward visioning cannot cease. All the

aphorisms of men are like unto "no royal road to learning," but according to the lore and law of divine mysticism, the road to the learning that falls on the face of the
upward watcher is royal. The upward watcher knows
things which before he knew not, and which neither
teachers nor books have mentioned.

The downward watcher wades through seas of trouble
and is chided for not having faith. How can he have
faith, the substance of things hoped for, when it is the
fourth smite from above, reaching down over his own
isolated vision to the roots of his own being, and
rousing his own untaught spark of authority over an
undescribable Almighty Executive?

When the spark of faith like lightning for splendor
spoke from the masterful lips of the Unkillable Redeemer, the quickening Mystery back of the tomb and the
soldiers' swords flung them all aside for His free feet to
go into Galilee, where the eyes of five hundred might
see Him alive and not dead.

Let us write it with a pen of light dipped in the
fountain of everlasting truth, that we have found a new
Servant—The Able-to-do all things. "Is there anything
too hard for me?" He saith. "Before the day was I am he,
and there is none that can deliver out of my hand. Concerning the work of my hands, command ye me." "Kings
exercise lordship," said Jesus. How shall one be king
except his kingship be roused? And true kingship, one
ray of which is as strong as the decree of Darius concerning the lions, comes from above: "By me kings
reign," saith the Great Voice that John turned to see.
If lions do not stand back, and warrings do not cease, and
diseases do not retire, the true kingship is not among
us. Only its crude symbol, working through the heavy

machinery of army and navy, and jailor and hospital faces us.

Then the disciples asked Him to increase their faith. But He answered them nothing. For faith, which is kingship exercising to call the God of Lazarus to come forth, and the God of the withered arm to appear, is the deepest secret of all the deep secrets of the *Magia Jesu Christi.*

Seeing then that after speaking with commanding earnestness to Unseeable Majesty He made insanity and poverty drop their grip, they said, "Lord, teach us also to pray." So He taught them to speak firmly and sternly to the Great Servant.

He that is greatest among you, let Him be your serv- ant. He that is greatest among us through all eternity is the Lord Strong and Mighty. None other is greatest. Your brains do not make your inborn Self greater than Self of your serf; your money cannot make you greater than the pauper; for God is no respecter of persons, and He surely hath made of one blood all the nations. Only One among us is greatest. Our Father is He—with name unspeakable on the lips of the downward watcher. His is the kingdom ready to show its finished presence, His is the will to command and the will to obey, identical with the inborn root of obedience and authority inherent in man, the highest God and inmost God being one God.

He feeds with super-substantial bread all who rise up like Marcella to demand it. He rouses the payment of the debt of confidence to command owing unto Him since ever the world caught us in its wheel. He delivers us at our bold command. He prevents our speech when it chooses the path of disease in description of evil. And this is His way and His glory, though you believe it not and take not hold of this key to His kingdom. "He shall feed thee on the Heights of Confidence," prophesied the exiled Ezekiel, writing by the banks of the Chebar.

The speech of the Lord's Formula being understood as the word of command, acts like nutriment to the hidden Jehovah nature. It is that feeding on strength which Micah was exalted to foreknow: "Man shall feed in the strength of the Lord." It is the end of that feeding on descriptions of goodness and badness, poverty and riches, the pairs of opposites, which Solomon noted as the foolishness that only fools feed upon.

As the savage can be taught mathematics and become proficient therein, centuries before he could evolve mathematics by himself, so are we taught the obedience of the Good Will fronting us by the gold formula of the Prayer of our Lordship long before we could have formulated it. "I know that his commandment (or the commandment of Him), is life everlasting," said the Messiah. "Thou through thy commandments hast made me wiser than mine enemies," cried the glad psalmist. For when I demanded that Thou "bow down thine ear to me (to) deliver me speedily, and be my strong rock," then "thy gentleness made me great."

"This is the whole duty of man," said Solomon in one of his moments of speaking above his mind, "Fear God, and keep his commandments."

"God manifests his word according to the commandment of God," wrote John the lover. "The commandment of the Lord is pure, enlightening the eyes," wrote one who had touched kingship by watching toward the high I AM who makes kings.

At the Waters of Lourdes, some patients are taught

to repeat the great formula of the hidden Lord in man, called the Lord's Prayer, fifteen times while the curing waters are being tasted. Has anybody explained to them that the waters of tribulation begin to subside for him who touches the fifteenth cubit above them? Has anybody explained that Paracelsus the miraculous healer of Zurich, caught all his flashes of genius from much repetition of the seven stately commands of Matthew's Lord's Prayer, or prayer of our own lordship? 1. Hallowed be Thy name. 2. Thy kingdom come. 3. Thy will be done. 4. Give me this day my super-substantial bread. 5. Forgive my debt of confidence to command Thee. 6. Let me not into temptation. 7. Deliver me from evil.

Who is not glad to utter these words of insistence, that he also may be healed of his cringing to old age, and death, disease, and poverty?

The fourth angel being caused to fly swiftly, smiteth all of them of the mystic formula. They are those that keep the commandments of God and the faith of Jesus, as John on Patmos foresaw. The fourth angel is the bright angel smiting Jacob, smiting Peter, smiting Job "on the left hand where he doth work."

Look unto Him of power to stablish according to the mystery which was kept secret since the world began, but now is made manifest according to the commandment of the everlasting God, for the obedience unto faith, as did Paul, writing to the Romans with the pen of that same Victorious Confidence, even to the quickening of dead Eutychus.

Faith is man's El Shaddai, his risen recognition of himself as Jehovah Soul, seeing the mystery of Divine Obedience everywhere awaiting the kingly rise of his heaven-planted boldness to command, "I will not let Thee go except Thou bless me!" "Thine is the kingdom, and the power, and the glory, forever and ever."

> "A Shape looked up from eating herb and grain
> It chanced to see the stars, and with that look
> Came Wonderment, and Longing in its train.
> The food untasted lay. A beating pain
> Smote at its forehead, but it looked again
> And yet again. And then it thought.
> Lo! Man stood upright as the stars did wane!"

V

WORKS

PROLOGUE

When the Greek and Roman peasantry cried aloud to invisible Mercury, "Grant us magistral to poverty!" they received magistral to poverty. Their confidence was effectual. Confidence, or faith in anything, makes it work according to the confidence. "Faith in a dead sardine's head's healing power would make it heal," declared the ancient Japanese.

Higher up in the scale of authorities there was Carlyle insisting that "conviction is not properly speaking conviction till it develops into action." Then there was Paul the Christian convert finding that "faith without works is dead"; by which he meant that unless some kind of work takes place we haven't believed anything.

Our globe has been called Number Five, the planet of works, since everybody and everything must accomplish something or be nobody and nothing: Sun, Vulcan, Mercury, Venus, Earth. Each to his feat, or opus, till some crowning bloom in earth's garden of man cries, "It is finished!"

Earth, as Number Five, must perform The Great

Achievement. The most wonderful achievements of mankind have been brought to pass by confidence in some wonder-working Unseen Power. Moses and Aaron had a five-pointed star at the end of their mystic wand which they swung high into some unseen perfect land. And when the workings of that land touched this earth they were called miracles: "And Moses stretched forth his rod toward heaven . . . and fire ran along upon the ground . . . and hail smote every herb of the field . . . only in the land of Goshen where the Children of Israel were, was there no hail . . . and Moses spread abroad his hands unto the Lord and the thunders and hail ceased"—even over all great Pharaoh's land.

All work is redemption. It redeems a place or a people or a situation from one status into another status. And redemption is historically associated with Number Five: With five loaves did Jesus redeem five thousand people from hunger. With five measures of parched corn was Nabal redeemed from death. With five measures of parched corn was Abigail redeemed from commmonplaceness to queenship. With five encounters with the terrible Archelaus did Heracles redeem the whole land from misery. With five sling stones did David redeem all Israel from Goliath the terrible. With five men did Joshua give his people rest from their enemies on the side of Jordan toward the sunrise. With five wounds did Jesus redeem common mankind from ignorance of His Sonship to Royalty Triumphant.

"If a man steal one ox, let him give five oxen for the one ox," and he shall be redeemed from the stigma of thief. He shall be restored to his former estate. He shall be reinstated. It shall be as if he had never stolen.

The sardonyx stone, which was the fifth stone of

character building according to John the Revelator, was worn by the ambassador of the King. To him was given power to redeem such as were appointed to destruction. We read that Joseph wore the sardonyx stone as Vicegerent for Pharaoh king of Egypt; that Haman wore the sardonyx stone as representative plenipotentiary for Ahasuerus king of Persia; that Philip wore the sardonyx stone as ambassador for Antiochus king of Syria, and that each of these had power to redeem such as were appointed to destruction.

It has even been traditioned that Jesus wore the sardonyx stone as Ambassador Plenipotentiary for the King of Kings and Lord of Lords, but there is no written history for this tradition as there is history for the lesser vicegerents of lesser kingdoms than the whole earth.

"Five truth mumblings are self active." Five eternal words were traditioned as written on the shining garment of Jesus the Glorified.

If we please to look up all that has been written about the mystery of *Five* we shall see how worthy is Number Five to be called *The Works Lesson*. It answers the Hindu discovery that the best doctrine is that which removes pleasure and grief from the mind; showing that doctrine is self active. So the Fifth, *Works,* must be a working doctrine, acting on the mind, which affects the body; which body is the working field of mind. Vision affects the mind. Mind is the working field of the vision, as body is the working field of mind.

Notice Hegel finding that "we always look toward an object before thinking it." Mind glorifies or cramps the body according as the visional sense runs high or low. Nine hundred ninety-nine thousand nine hundred ninety-nine diseases and pains were declared by the

Parsees as having been formulated by low visioning acting on mind to afflict its body.

The best ambassador for any king would be he who
should best understand his king's mind and best carry
out his king's hidden wishes. There is One King of
Kings and Lord of Lords, whose whole purpose toward
His kingdom has ever been peace, health, wisdom,
majesty even to the greatness and wisdom of His Own
Self: "Look unto Me." "I extend peace like a river."
"I am the Lord that healeth thee." "I will instruct thee
and teach thee." These are the words of the High Redeemer inhabiting Eternity, whose way upon the earth
is the saving health of the nations. The Ambassador
Plenipotentiary for this High Redeemer said that the
same fountain sendeth not forth both bitter and sweet.
So when we have pain or poverty or sickness or misfortune of any kind we must have been looking away
from the High Redeemer, who hath counselled, "Seek
ye Me and ye shall live."

Let us not be deceived by the poetic eloquence of
any great poet or theologian who tells us that the King of
Kings suffers or is grieved. For if He suffers or is grieved
He must shed suffering and grief around Him, even as
we when suffering and grieved shed suffering and grief
in all directions. Those who speak of the King of Kings
are not ambassadors understanding the Great King's
mind when they report that He is angry with the wicked
every day, or that anything grieves or dismays Him. They
are ambassadors for their own kind of king, and their
own kind of king works his own way with them.

Notice that high potentate who spoke of Jesus Christ
as a fable. Bespeaking no balm of Gilead in this "Fable,"
the potentate was afflicted with an incurable malady;

his fortune melted; friends failed him; his great ambitions faded on all sides. For there are some wounds on life's pathway that only the Real Christ Jesus can heal.

Hear the Erythrean Sibyl prophesying of the Real Jesus Christ seven hundred years before He appeared: "All who style Him King shall be happy in His Kingdom." Read the words of Zoroaster òf Persia eighteen hundred years before the coming of the Anointed of the Heavenly King: "A virgin shall conceive and bear a son, and a star shall appear at midday to signalize the occurence. When you behold the star, follow it whithersoever it leads you. Adore the mysterious Child, offering Him gifts with profound humility. He is indeed the Almighty Word. He is indeed your Lord and everlasting King."

There is a science that runs like a river of light above all the sciences. It never changes its assurances. It tells of the Working Efficiency of One Lord Supreme and of how the language runs that describes the Working Efficiency.

It is the Mystical Science. According to its practice we never say we fight for the Lord Supreme, but "The Lord shall fight for us and we shall hold our peace." We never say we trust a friend, but "All my trust on Thee is stayed."

For, "Put not your trust in princes," is the language of the Lord Supreme; "I will contend with him that contendeth with thee"; "No man shall set on thee to hurt thee." "Fear not, I will help thee." "Look unto Me."

Mystical Science starts the New Language promised sometime to break forth from the lips of mankind: "I will turn to the people a pure language." "They shall speak with new tongues."

Whatever language mankind starts with shows that
visional sense has preceded speech.

Man can throw his vision out toward damage for
some neighbor and silently mentioning the damage in
definite terms he will find it formulated in the experience of that neighbor. But "Add to your strength, knowledge," said Peter. Is it not written that the man who
imagined the damage of his neighbor, fell and injured
the very limb he had imagined himself using with neigh-
bor-damaging violence? David found out this law: "How
long will ye imagine mischief against a man? Ye shall be
slain all of you; as a bowing wall shall ye be, and as a
tottering fence."

See how we need high watch with its high language!

People must learn the law of lifting up the face to
the Lord Supreme who worketh noble conditions of life
into view.

How shall we make it plain that power and vigor
and plenty hail from above, with no need to maltreat
or suborn our neighbor?

Things and people are often troublesome. Mystical
Science teaches us to let them alone as if they did not
exist, and look up to the Vast Vast Countenance for one
second of time; maybe two seconds; to have nothing
to do with them; to cut the threads of attention toward
them. The Vast Vast Countenance saith, "I will set them
in order before thine eyes," "I restore to you the years
that the locust hath eaten." Have not locusts always been
symbolic of domestic tormentings? Restorations hail from
above. Expect greatly from above, and greatly shall restorations multiply. "Prove me now herewith, saith the
Lord of hosts, if I will not pour you out a blessing that
there shall not be room enough to receive it."

So is Self-recognition awakened. So is new mind built. So is hidden ability set astir. So ariseth the new race of which Jesus was the forerunner. No man has ever stood so boldly forth for the Redemption of the God Self of man from the clutches of the mortality self as that young man of despised old Nazareth nineteen hundred years ago! No lover of his brother man so willing to die that He might show man his own God transcendence has ever appeared on this earth! He knew the ancient doctrine preached in many ways that man was the off-spring of Satan with only one God glow in his being, and that the angels dwelling in glorious Paradise had asked each other who was willing to leave his heavenly home to redeem the God of man from the Satan of man and daringly declare to man, "For this cause came I forth into the world." Why should not now the angels,

> Sing oftentime the story
> On heights of untold glory
> Of the greatest one among them,
> Christ Jesus and His love.

This Study is self-active in its treatment power. Do not try to delve into it; make the acknowledgments and let it have its Apostolic way with you.

E. C. H.

V

WORKS

It was practicing inborn authority over the Universal Servitor, when the wonderful Jesus cried, "Glorify thou me!" And when on the cross He acknowledged, "How thou hast glorified me," He was seeing the obedience of the obedient God to His orders. His eye overlooked future ages, when He should stand to mankind as the embodiment of divine insistence—His Name above principalities and powers, and above every name that could be named.

Abraham was practicing inborn authority over the Invisible Servitor, when he said, "I lift up mine hand unto the Lord, the most high God, the possessor of heaven and earth.—Whereby shall I know that I inherit the land?" And he was experiencing the obedience of God when a deep sleep fell upon him, and he saw himself famed for spiritual and material riches throughout all succeeding generations.

David was practicing innate authority over the Universal Obedience, when he said, "Be thou my strong rock! Deliver me speedily!" And when David's little band of warriors had swelled into "a great host, like the host of God," and he had been three times crowned king, he was in the thick of God's obedience.

Gideon was practicing the same inborn authority, when he spoke to the Universal Servitor, "Show me a sign that thou talkest with me." And when fires rose up out of a rock, and the Midianites and Amalekites and all the children of the East fell down at the sight of Gideon, then the Great Servant was obeying Gideon's bold prayer of lordship.

Authority with the Universal Servant is roused in us, as in Gideon, to accomplishing vigor, by repeating the prayer of our inborn lordship, with firmness and sternness. (See Fourth Study.)

Authority with the universally present Divine Servant discloses authority with the particularly present divine Self, or Spirit of God vivifying each frame. "He that ruleth his own spirit is better than he that taketh a city," wrote Solomon the discoverer of spiritual activities. This individually present Godship is that Self which obeyed Kossuth, when he addressed his own body, seeing it as charged with an intelligent entity, saying to it whenever it fell into weakness, "Rise up strong and active; be competent to do all my work this day; throw aside pain!" It was the God charging every molecule and atom of him with competent obedience that slowly stirred from its quiescence into energetic activity, intelligently obeying his orders, making him strong and healthy for the day.

"The self of the man who is self subdued is as the Supreme Self, or God," wrote the Theophoroi of old. The Eternal Immanence is in the present tense exactly as in the past, and the still intelligence that waits at every infinitesimal pore of our human frame, today, as yesterday, leaps into action if we command with firmness and sweet sternness. So now let us according to orders, "Upraise the self by the Self; for Self is the friend of self."

Every night of the life, before the eyes close in slumber, the immanent Godship swelling the self with possibilities, should be commanded what work we choose ourself to accomplish, and what type of character we choose to exhibit. "Awake up, my glory!" commanded David, and his glory awoke. "Shake thyself from the dust; arise, and sit down, O Jerusalem; loose thyself from the bands of thy neck, O captive daughter of Zion!" shouted Isaiah to himself. And Isaiah transcended all the prophets that have ever lived on the earth; he was loosed from all dependence on the instructions of mankind.

The Self of ourself has a voice. Its answer is, "I can all that and more," to every command we give it. Why should we fall asleep regretting the day or dreading the morrow, when we have an eternally abiding Self, quiescent, still, instinct with executiveness, waiting our firm insistence on Its action in our behalf?

Apollonius realized that he must be up and about the business of managing his own Godness, and he commanded the vitality of his own mind to remember all things, and the vigor of his own heart to beat with steady hardihood. And his mind did remember all things, even its relation to Universal Spirit that raises the dead; and his heart beat so in rhythm with the universal possibilities of himself that he was able to be in two places at once, whenever it was necessary.

"The upright shall have dominion," chanted the Hebrew choirs under the leadership of David.

"Praise is comely for the upright," sang the same Hebrew choirs; for the "I," the Soul-Self, the God-Self is one with praise as with command, ready to demonstrate all excellence for which It is praised, as all accomplishings to which It is commanded. The voice of inspiration teaches us to praise Soul, the upright Self charging ourself like a Shekinah pillar of fire by night, and a straight cloud of glory by day.

Let us not speak dispraisefully of our "I," our secret free Spirit, saying, "I am sick," or, "I am discouraged," or "I am inconsequent"—for this is speaking out of key with high truth. "I was free born," said Paul, speaking in key with truth. So were we all free born, upheld by Free Spirit forever. By recognizing this we bring It to the front.

Lysias said, "With a great sum obtained I this freedom." He had had to struggle to appreciate his free born "I." Paul saw that no strenuousness is called for; that truth is mighty in itself, and whoever fights for high truth has forgotten its almightiness. Lysias represents those of us struggling to sense our freedom. Paul represents those of us sensing our free estate by simple recognition of our free Divinity.

"Lo, my sheaf . . . stood upright," praised Joseph, visioning ahead when the Jews of all ages should owe their daily bread to his fidelity to praise of his own Self, maintaining his own vision of himself, even while in prison, as one instinct with majesty and virile with omnipotence. He had once had his ordinary senses entranced as in a dream, while his sense of Godness rose like an incense, and, remembering this, he told his fellow prisoners the import of their dreams, while yet his own dream was unfulfilled. He knew that truth is truth whether we are as yet embodiments of it or not.

Poets by setting their words and thoughts into tune with the soundless whispers of their own laws of Soul life, have often struck the chords of Self praise like wonderful antiphons:

"Thou shalt flourish in immortal youth,
  Unhurt amid the war of elements,
  The wreck of matter and the crush of worlds."
"Wingless upon your pinions forth I fly,
  My words begin to breathe upon your breath:
  Shorten half way my road to heaven from earth."
"It were a vain endeavor, though I should gaze forever
  At that green light that lingers in the west;
  I may not hope from outward forms to win
  The passion and the life whose fountains are within."

The muscles of the body can be trained to be so strong that they can beat down giants in pugilistic encounters. Thoughts of the mind can be trained to be so strong they can strike down opposing ideas on invisible mind fields, and paralyze the judgments of the brain so that judges and juries speak nothing, or, speaking, speak only nonsense. Muscles highly trained have won out against natural muscles, and thoughts highly practiced have wrought mental havoc. What shall we, who wish to be free and not to engage in warfare, do, when our peace and safety are menaced by foes of such giant physical and mental stature? We will seek unto God, the High Presence in the universe not affected by thoughts: we will seek unto Him present at our own headquarters, Whose years alter not, Who saith,

"These all shall perish but I, Soul, Self, shall endure."
"Seek ye Me, and ye shall live."
"No weapon formed against thee shall prosper."

It shall not profit a man to gain the whole world by the prowess of his arm or the might of his thought. It shall only profit him to know his own Soul, uncontaminated offspring of Eternal Majesty, whose triumphs are already complete, ready to manifest.

Highest God and inmost God is One God.

Let mind no longer claim creative powers or accomplishing energies. The true work is already complete
in Spirit, the Self that we praise. "I am all that and more,"
answers Soul, our own "I," to our highest descriptions.
"I can all that and more," It answers to our highest man-
dates.

Soul hath a strength of empire and an influential
glory which it hath not entered into the heart of man to
conceive:

> "Thou hast
> A strength of empire fix't,
> Conterminate with God."

Our own Soul, our own free Spirit forever says, in
bold faith, "I am Truth, I am God—Omnipresence,
Omnipotence, Omniscience." The outer appearance, the
cocoon, the hard chrysalis, vibrates when the words of
Immortal Soul are spoken silently or audibly, as the
chandelier hums when its key note is struck, or as the
brim of a bell resounds when its hidden tongue hits it.

Paganini said he could shake any building by maintaining the note that caused it to vibrate. By speaking the
truth of and to our own Soul, or Self, we strike the true
key tone to the body of flesh and its mind and emotions.
We can speak in silent language or audible words the
truths of the transcendental Self that cause health, happiness, and helpfulness to radiate; and this speech wakes
the Soul type of man to walk on earth.

A great modern philosopher says, "The Spirit is ever
rising up in wrath against the forces that would brutalize
it; the Soul is ever striving for independence of matter."
But the truth of Soul is, that It ever dwells in calm

majesty, striving against nothing. The heathen philosopher spoke more wisely: "Nothing can injure the immortal principle of the soul."

"Truly my soul waiteth upon God," sang David to a noble tune on a stringed instrument—"Truly my soul waiteth on God." And the angry javelin of the angry Saul could not reach him, for he had keyed himself to uninjurable Immortal Soul, by voicing Its invigorating truth.

It is not what happens to us that makes us healthy, happy, radiantly helpful; it is what we harmonize with, and we harmonize with what we describe. The winds of misfortune and pain hit every one sooner or later; but they do not touch our Self. "Truly my soul waiteth upon God," we sing. "My soul doth magnify the Lord," we chant. We set ourself to the heavenly Soul key by praising our innate Lordship, our eternal identification with Divinity:

> "'Tis the set of the soul
> That decides the goal,
> And not the storms of life."

The Hopi Indian praises the great power that shines behind the sun. The Parsee praises the great power whose shining creates the sun. We praise the great free Spirit that stands back of our mind, which mind was once to us the sun of our life. We praise the free Spirit that knows beyond the mind, which is saying ever, "I am God—I am Truth—I am Light"—and so we touch the law of the *five*. For when the fifth angel sounds, the sun and air are darkened to the vision of John the Revelator. He means, that by recognizing our divine, "I," our mind ceases to be our supreme guide, and the sensations are forgotten.

There is a consciousness of cold, there is a consciousness of heat, there is a consciousness of stinging, and of falling or rising; so there is a consciousness of God. It is consciousness as in a trance; and John calls this the darkening of the sun and the air; for John is always speaking in figures.

He who has the consciousness of God knows beyond his mind, and wakes a new kind of body, in tune with the Infinite Immortal, the Lord Supreme. Nicholas of Basle once had this consciousness transcending his mind suddenly spring forth through all his being, and he found himself for a moment one with the Origin of knowing; and he said strange things beyond his previous concepts.

Praise and command of the divine Self of ourself always wakes the consciousness of our own superiority to environing disadvantages and ignorances. "Know Thyself" was written over the Delphic Temple. It is only the divine Self, Soul, free Spirit that is worth knowing, worth praising, worth commanding. Fight as though thou wert the fighter, but know that it is the free Spirit of thee that moves on the opposing phalanxes that try to make life difficult—and the free Spirit masters them though the mind and the flesh quake. Jacob trembled all night by the Jabbok brook, and his mind was afraid, but the Angel of God's presence, with whom he had identified himself, fought his battle for him. "I have seen God face to face," he said, "and my life is preserved." "The Angel which redeemed me from all evil, bless the lads."

In the Fourth Study we are taught to practice the prayer of our lordship. John the Revelator calls this the smoke of the incense arising from the pit of our own infinite possibilities, for the Highest and the Inmost are one in infinitude. "And the smoke of the incense, which came with the prayers of the saints, ascended up before God out of the angel's hand." And One came like a star from the skies, showing to all mankind their own infinite possibilities, through recognizing the identity of the Soul of each man with the majesty of Almighty God. The infinite possibilities of Soul, our God Self, are spoken of in the Apocalypse as "the bottomless pit."

When Naomi recognized Ruth as her leader and guide, and the light of her life, she was recognizing Spirit, and forgetting her unhappy mind and emotions. "Entreat me not to leave thee," said Ruth "nor from following after thee, for where thou goest I will go." And Naomi, by her acceptance of Ruth became forebear of Christ the Saviour. So our Soul, our Ruth, is ever saying to us, "Where thou goest, I will go," through the ages—always waiting in quiescence, in shining *esse,* for acknowledgment by praise and command.

"You cannot praise Me so highly that I am not more than you praise, you cannot command Me so greatly that I cannot work by you still more greatly," ever whispers our secret Self.

The young Jesus stood up in old Nazareth and spoke of the everlasting Son of God, the Immortal Youth, the Unconquerable Divinity of man, and the Nazarenes tried to throw Him off a precipice, to destroy Him and His words. They refused to recognize their own divinity, their self renewing fountain of immortality. So old Naomi, accepting what Nazareth rejected, was prophecy that the seed of the woman should bruise the serpent's head—or, that the vision of woman toward the ever fresh fountain of divinity should save the world from the hard rulership of mind and matter.

the mind no longer conceives itself to be the knower, recognizing that Free Spirit is the knower and the doer, then is man's liberation from the laws of mind and matter," intoned an illuminated Hindu sage.

It was by the recognition of His own Infinite Divinity, His own Godness, that Jesus of Nazareth discovered His ability to perform the greatest work ever accomplished upon this earth, and made Himself the Bloom in the Garden of Man of all efforts to accomplish great help-fulness by divine at-one-ment. He saw Himself as the ful-fillment of the prophecies of the ages, that one should come who should be greater than death and pain and grief and all the hatred of all the human race. He saw himself so identified in the flesh with flawless, unhurtable Substance, that He could take to Himself all the pains and discords of the human race, and yet be not slain, and yet be nothing less than Divinity. He saw that who-ever should in future ages acknowledge His accomplish-ment, should be set free from his own pains and discords, and should sense that Jesus of Nazareth, charged with His own divinity, was the Saviour of the world from disease and death, misfortune and decay, even here and now upon this earth, in the sight of all mankind. "Who gave himself for us," said Paul, "that he might deliver us from this present evil world."

"Who hath believed our report? and to whom is the arm of the Lord revealed?" wrote Isaiah, visioning ahead when the Saviour of the world, as the arm of the Lord revealed, should not be acknowledged as having de-stroyed death by taking into His own body the sum total of death, and not acknowledged as having destroyed disease and pain by having taken into His Divinity-charged body the sum total of disease and pain, that all mankind might go free by the acknowledgment. Yet, "Surely he hath borne our griefs and carried our sorrows—" that we might go free from grief and sorrow. "He *was* wounded for our transgression, he *was* bruised for our iniquities; the chastisement of our peace was upon him; and with his stripes we are healed."

When Jesus came, the Bloom in the Garden of Man, of all the divine doctrines of earth, and charged Himself to the complete with the Divine Presence in the universe, He fulfilled the prophecy of the Jews that one should come who should be so at one with Absolute God that He could be slain and yet not dead, and diseased and yet immaculate, who should chemicalize out of existence, and thus make nothing, all the maladies of earth. The condition of other men being consciously and visibly saved by this exercise of His divinity should forever be, *the acknowledgment of this accomplishment,* "He that acknowledgeth the Son hath the Father also."

"When thou shalt make his soul the offering for sin . . . the pleasure of the Lord shall prosper in his hand."

What is the pleasure of the Lord?

"It is your Father's good pleasure to give you the kingdom."

As we each of us have a work which is supremely ours and no other can accomplish this opus, or God-ordained work, save our own self, so Jesus of Nazareth had His work, and His work was the redemption of mankind from sin, sickness and death, by the withdrawal into Himself, by virtue of His supernal Godness, all the sin, sickness and death of the universe, leaving the universe entirely without sin, sickness and death, thus making us to walk through a redeemed world.

This was His chosen work; and as it would be only

fair for mankind to acknowledge the fact, if we had composed the greatest piece of music, or had built the most splendid temple, or discovered a wonderful law of mathematics, so it is only fair to Him, Jesus of Nazareth, charged to the supreme with Christ power, to acknowledge the completeness and splendor of His finished chosen work.

The inspired Scriptures are explicit on the subject of His successfully accomplishing in the large, what Elisha and Elijah accomplished in the small, in the way of taking death into Himself, that He might deliver them who through the expectation of death were all their lifetime subject to its bondage, destroying him that hath the power of death, that is, the devil, the lie from the beginning, abolishing death once for all. "Who hath saved us, and called us with an holy calling, not according to our works—who hath abolished death."

Gautama Buddha offered to bear the sins and the consequences thereof, which the Kali Yuga, or age of spiritual blindness was bearing, but he could not accomplish it, because he had not sufficiently identified himself with Divinity Supreme.

Elijah took the death of the Zarephath woman's child into himself, and because he was so alive, so virile with spiritual fire, he chemicalized death into non-existence for the child.

Elisha took the death of the Shunamite child into himself, and because he was so strong and alive with the Spirit of God, he chemicalized the death of the child into non-existence.

It is a well-known law in some countries that certain people by putting themselves into certain attitudes of mind and sensation, which we would call the conscious-

ness of God, can withdraw sickness and pain and disease
and deformity and death into themselves, leaving the
sufferer free from his sufferings. Manes, of the Man-
ichean sect of Christians thus took the sufferings of a
great many people into himself, leaving them free from
suffering.

Catherine of Siena withdrew many diseases and other
forms of affliction from unhappy victims, into her own
self, and by virtue of her spirituality made them nothing
for them, and for herself also.

John Joseph of Cocenza, a small city southeast of
Naples, received into his own body the two terrible
ulcers with which the Archbishop Michael was afflicted,
and made them nothing by the chemical action of his
awakened spiritual Substance.

This vicarious suffering is often taken in our own
day, by sensitive and spiritually illuminated men and
women, who are not awake enough to chemicalize the
condition into nothingness, and so every neighbor wonders why spiritually sensitive and divinely illuminated
people are so often afflicted in mysterious ways.

It is only Jesus of Nazareth in the history of man,
who has understood how to consciously withdraw the
wretchedness of the people into Himself, and make
wretchedness nothing both for them and for Himself.
He did it by the consciousness of His own God Substance,
His own majestic, untarnishable Soul. And Isaiah the
prophet, gave the assurance that all should go free from
their own sorrows and sicknesses, who should acknowledge that Jesus of Nazareth, by the Soul, or the Christ
splendor shining through Him, had borne their griefs
and carried their sorrows, taken their infirmities and
borne their sicknesses.

Mistakes of mind and action may be conscious or unconscious on the part of mankind. When mistakes are unconscious people may never trace the mechanical consequences of their mistakes in the misfortunes of daily life. The mother compels the child to study his lessons, not knowing that his eyes or brain may be weak, and in after years, when he is insane or blind, she is totally unconscious that she had once pressed his brain or eyes beyond their bearing point. The father compels his child to eat food repugnant to it, not sensing that his offspring naturally divines its proper pabulum, and when later the child has scrofula or consumption, the father feels that it is a great affliction, but surmises not at all his own guiltiness.

"Art thou come hither to destroy my son, and call my sin to remembrance?" asked the weeping Zarephath woman, for she belonged to a people and an age of the world which believed that it is someone's mistake of mind or action, exhibited in result, when sickness or death or deformity or misfortune attacks anybody.

Keshub Chunder Sen of the Brahma Samaj of India, told in England, that the mothers of India try to direct the sicknesses of their children into their own bodies, by gashing and striking themselves, feeling that they are strong enough to bear the consequences of their own mistakes, which the Hebrews would call sins, quite certain that by this vicarious suffering the children are set free from pains and sickness.

It is a law which has slipped the attention of the scientific men of our age, with the exception, perhaps, of a very few. Charcot of Paris discovered in the hospitals, while practicing healing by suggestion, that the malady of one person set free by his hypnotic will was found lodged in the next ward in some other patient, or even in the body of some person mentioned by his liberated patient.

Sometimes mental practitioners of today wake up in the morning with the pain of the person they have so faithfully treated, silently, with the noble affirmation, "You are Free Spirit, uncontaminated by disease or sickness." If they are spiritually strong and healthy they throw off this vicarious bearing, but if they are not, they may keep this condition for some unpleasant length of time.

A certain New England healer of this sensitive type did not like to shake hands with sick people, because she caught their diseases, though they went free, and she was not aware of being spiritually strong enough to dispose of the condition she was thus vicariously bearing, and not acquainted enough with the offer of the Great Vicarious to acknowledge, "Himself took our infirmities and bare our sicknesses," and so pass it along to its annulment.

This is where the offer of Jesus the Christ—the vicarious bearer of all the sufferings and all the unwitting causes of sufferings, for all the world—the destroyer of karmic death and disease, should be acknowledged, which acknowledgment is the same as passing on all distress to the Healing Fountain: "Himself took our infirmities and bare our sicknesses"—and "redeemed us from the curse of the law."

There is no describing what world-wide liberation from suffering might be manifested, by making "His Soul an offering for sin" and its consequences, since the sacred promise remains on eternal pages, "That the God of our Lord Jesus Christ, the Father of glory may give unto you the spirit of wisdom and revelation in the knowledge of him."

Peter and Paul among the early Apostles of the Christian doctrine were the most definite and distinct in proclamations concerning the mysterious mission of the Lord Jesus of Nazareth.

To the Galatians Paul wrote, "God sent forth his Son, made of a woman, made under the law, to redeem them that were under the law." "Christ hath redeemed us from the curse of the law, being made a curse for us."

And to the Corinthians he wrote, "God was in Christ, reconciling the world unto himself, not imputing their trespasses unto them."

And to Titus he wrote, "Who gave himself for us, that he might redeem us from all iniquity."

In calling the attention of the Hebrew Christians to the majesty of the fulfillment of the law in the history of the Redeemer, he said, "We are sanctified through the offering of the body of Jesus Christ once for all." "How shall we escape if we neglect so great salvation?"

He further explained to the courageous Christian preacher to the Cretans—who were the famous liars of Homer's time and of Paul's time—that Jesus had given Himself for us all that He might redeem us from all iniquity.

It was reiterating the insistence of Jesus Himself that the "Son of man came to give his life a ransom for many," when Paul wrote so boldly of Him "who gave himself for our sins, that he might deliver us from this present evil world," "Who though he was rich, yet for (our) sakes he became poor, that (we) through his poverty might be rich."

Paul does not try to argue us into the acceptance of

the principles of vicarious suffering, and the liberation from suffering, accomplished by acknowledging who hath suffered vicariously, for Paul was a Jew of the strictest sect of the Pharisees, and it was to him the natural religion that one should be made sufferer for the transgressions of many.

David had slain the seven sons of Saul to stop the three years of famine in all Palestine, and no Jew of Paul's time doubted that the cessation of the famine in Palestine in David's time was accomplished by the vicarious suffering of the seven sons of Saul; and Paul had been brought up on the belief that the transgressions of many might, even as a religious sacrament, be solemnly passed on to some innocent animal in the wilderness.

He understood Jeremiah's lamentation that great and small in his time were dying in the land, because of the sins of their parents, their prophets and their priests; and Paul understood also the unbelief of the scientific Greeks, and the hesitation of the Jews in accepting the vicarious accomplishment of the Universal Redeemer, because he knew that although it is a strict law of possibility, it is a subtle proceeding, and only the mystically visioned can truly see it in the world-wide as the old Jews had seen it in the national.

"For the preaching of the cross is to them that perish, foolishness," he wrote to the Corinthians, "but unto us which are saved it is the power of God."

When we are told, every one of us, to make acknowledgment of our works before the Heavenly Father, it is a severe test of our knowledge of Scriptural information, for there we are told that no man has ever accomplished any work purely by the recognition of his own divinity, except Jesus of Nazareth, who hath "put

all things under him . . . . to redeem us from the curse
of the law. . . . being the first fruits of them that slept,
. . . abolishing death."

Peter's words are vivid and emphatic like Paul's:
"Who his own self bare our sins, in his own body on the
tree, that we, being dead to sin, should live unto righteousness, by whose stripes we are healed," for "Christ
hath suffered for us in the flesh," and "once suffered for
sins, that he might bring us to God."

We are all posited on this planet for the one purpose
of accomplishing some great opus, or work of a unique
and inimitable sort, by the recognition of our own
divinity, our own free bold spirit, offspring of Almighty
Jehovah, and we have the example among the multitudinous sons of men, of One who was the first fruits
of them that slept in non-recognition of their own divine
equipment.

"For now is Christ risen from the dead," "For both
he that sanctifieth and they who are sanctified are all
of one."

There is no respect of persons with God, and though
Jesus of Nazareth was indeed the first to accomplish the
superhuman, by the recognition of His own superhuman
equipment, there is no reason why each one of us should
not rise up and accomplish the super-mission which we
came here to accomplish for the glory of our Father
Eternal.

This is the great planet of achievement. Every individual upon it naturally seeks to accomplish some beautiful and praise-worthy deed. If we name the sun as the
first globe of our constellation, we are the fifth of the
round balls of our enfolding skies, and may easily be

called the planet of works, achievements, accomplish-ings, labors: *Sun, Vulcan, Mercury, Venus, Earth.*

But although we are the planet of bestirrings, we must remember and acknowledge that only One of us has really accomplished His native deed of unspeakable splendor by full recognition of His own unspeakable, splendid Divinity. Tauler the mystic was persecuted by the Beghards and Beguines, and grand convent ladies, for telling them that the only work they could acceptably present before the Majesty on high was the Finished Work of the Lord Jesus.

The early Christians associated Number Five with the vicarious sufferings and death of the first human being among us to recognize Himself as God:

> "Come tell me truly, to what truth
>   Should number five be guide?
> The wounds of Christ in hands and feet,
>   And in His bleeding side."

When David, who had read in Leviticus 20, that five should slay Goliath, he chose him five small stones, for he was about to accomplish the liberation of his people, a tremendous deed, by the recognition that his divine Self was coming up against an hundredfold powerful foe: "Thou comest unto me with a sword, and with a spear, and with a shield, but I come to thee in the name of the Lord of hosts."

Only David's secret Self, or heavenly intelligence, knew the name of the Lord of hosts, but by recognizing that somewhat about himself was Sonship to Jehovah, he took his symbolic five stones of victorious accomplish-ment, and undid the heavy burdens, and let the oppressed go free. Had David not acknowledged the victory already inherent in the unspoken Name of the Lord of hosts, Goliath, the embodiment of bondage would not have been abolished.

There is a divine executiveness accompanying all high acknowledgments, as Paul also understood, when writing to the magic loving Ephesians: "God . . . the Father . . . may give unto you the spirit of wisdom and revelation, in the knowledge of him."

As it has taken the light of some splendid stars thousands of years to reach our earth, so it has taken the best part of two thousand years for mankind to recognize the far reaching glory of the undertaking of the Divinity-charged Jesus of Nazareth. It takes the upward fling of all man's cognizance of the law of vicarious suffering, to make him at this late day give honor to whom honor is due, and proclaim to the High and Lofty One inhabiting Eternity—"Christ Jesus as Emmanuel, or God with us, has borne my mistaken actions and their consequences once, for me, that I might be unloaded of my life blunders and be free to accomplish my own great task.

"Christ Jesus as Emmanuel, or God with us has once taken to Himself the mistaken thoughts of my mind and their consequences that I might be unveiled of my mind, and free with my bold Soul, my uncovered free Spirit, to speak new words of victorious truth.—

"Christ Jesus as Emmanuel, or God with us, has once borne for me the burden of my human lot, that I might be unburdened free Spirit forever."

"He is the propitiation for our sins, and not for ours only, but also for the sins of the whole world." "That the saying of Esaias the prophet might be fulfilled . . . Himself took our infirmities and bare our sicknesses,—

. . . hath borne our griefs and carried our sorrows—" that we might not bear them.

All the learnedness of the world cannot compass the wonder of the mind of Christ, who knew all things and needed not that any man should teach Him. And the acme of His wisdom was His understanding of how to be God, glowing and transfiguring through the flesh, even to the annulment of all its moral liability; and how to transfigure the mind with new light, so that no more errors could darken it; and how to be so mighty with Omnipotence that all who should recognize Him should share His mightiness. Whoever confesseth that Christ hath thus actually once come through the flesh, partakes of the coming, and is himself sent as a worker of new work, and a speaker of new words.

Who is ready, by acknowledgment, to wash sometimes in this pool of Siloam, or complete negation of himself, in heavenly abandon to the great Scripturally proclaimed Neutral to the sin of the world, living, like Paul, only as "Christ (that) liveth in me?"

This is becoming dead with Christ that we may live with Him. "For if we be dead with Christ, we shall also live with him."

It makes a great difference to us what doctrine we mine out of the Scriptures, and the Apostolic Christians surely minded there from the doctrine that Jesus blotted out the handwriting of ordinances against us, taking it out of the way, nailing it to His cross.

Abraham was the pivotal man of sustained faith in individual good due from the Universal Absolute. Antisthenes was the pivotal man of cynicism, the doctrine of the absolute responsibility of the individual as a moral unit. Zeno was the pivotal man of stoicism, the practice of stern personal virtue as followed by Aurelius and Epictetus. Jesus was the pivotal man of achievement of the humanly impossible by the identification of Himself with the divinely possible.

The Chaldeans had prophesied of Him as the Lofty One to arrive among men. The Egyptians had foreseen Him as the Lord of the whole world to come among us. The Chinese had waited for Him as the Saving One to be born and die for the race. The Hebrews had expected Him as darkness expects light. The Sibyls had foretold that as a Saviour of man from his ungodliness should a free denizen of heaven come to earth to teach mankind of their own God Nature, and so redeem it from its hiding place in earthliness. "But," wrote the Erythrean Sibyl seven hundred and fifty years B. C., "hostile man shall spit upon Him; on His sacred back they shall strike; gall and vinegar shall they give Him to drink; on a tree they shall hang Him; a darkness of three hours from midday shall cover the earth. But on the third day He shall rise in joyful light, and all who acknowledge Him king shall be happy in His kingdom."

It is ours to let ourselves go bathe in the Siloam waters of yielding, under Scriptural orders, as meekly as the blind man bathed in Siloam of old, declaring the Scriptural doctrine of Christ Jesus as the first Divinity-awakened laborer in the vineyard, redeeming the world from sin, sickness, and death, that we might perceive with open eyes that we are walking through a finished kingdom, the arm of the Lord revealed. This is fasting from our sense of obligation to heal the sick, cast out demonic tempers, and raise the dead, since in Christ Jesus these works are finished, awaiting only acknowledgment to be plainly visible.

"Is not this the fast that I have chosen," saith the Lord, "to loose the bands of wickedness, to undo the heavy burdens, and to let the oppressed go free, that ye break every yoke?"

Let us go daringly free into Scriptural declarations, as this is pure Scriptural doctrine of vicarious bearing, mentioned distinctly to an age lost to all belief in the truth of the cross, though "all the light of sacred story gathers round its head sublime."

Great signs shall follow them that believe in the Redemptive labor of the Divinity-awakened Lord of Galilee. They shall disclose the heavenly health of the redeemed world their opened eyes descry—the world free from sin, sickness, death, misfortune, mourning. "For there is a kingdom on this earth, though not of it," said Balthazar, "that is a fact, as our hearts are facts, and we journey through this kingdom from birth to death without seeing it; nor shall any man see it till he first knoweth his own Soul."

Speaketh Isaiah of his own Divinity-charged soul, or of some other man's? asked the eunuch of Philip. Then preached Philip unto him, *Jesus of Nazareth*, the Deific Man of the first undoing of the shadowy mechanicals of human association, leading on the generations into new gerenda, or new works to shine forth with, through sighting with mystically opened eyes the Kingdom of Love and Life immortal and uncontaminate that lies all about us.

"When thou shalt make his Soul an offering . . . the pleasure of the Lord shall prosper." . . . "For it is your Father's good pleasure to give you the kingdom."

This Sixth Study is a lesson in invocation. It is well to know what we are invoking. Whatsover we desire we are invoking it, and sooner or later it arrives upon us. Let a man draw himself to himself, and leaving all else choose what he will, and so it shall be unto him.

E. C. H.

VI

UNDERSTANDING

As a bar of iron gains magnetic virtue by being placed for a time in a special position, so, it has been opined, particles of matter arranged and long continued in a certain posture may eventually become surcharged with life.

As natural phenomena are mysteriously symbolic of mystical realities we can understand by the bar of iron and the particles of matter how it came to pass that the inspirationally taught Theophoroi of old discovered that the inner vision of man ofttime giving attention to "Ain Soph the Great Countenance of the Absolute" would charge man with transfiguring newness of Divine Life in plain manifestation.

Great promises have been caught by awakening watchers toward the Vast Vast Countenance of Him who exalteth by his power, none teaching like Him:

"Look unto me and live." "Seek ye my face and live." "I give life to the faint." "And I will reveal unto them the abundance of peace and truth."

Take a wide, wide view of The Vast-Vast. "He giveth to all men liberally and upbraideth not." Upbraiding and stinting hail not from him, "not life but the Cause that life is—not spirit but the Cause that spirit is."

The oldest science in the world is the Mystical Science. It came to primordial man from the Tao over the Tao, while wistfully uplooking toward the wooing heights for some direction on his seemingly inscrutable life path.

Whatever came to primordial man concerning his relationship to the High Eternal First Cause he called *inspiration,* as it seemed to come along with his inward breathing. "The inspiration of the Almighty giveth them understanding," he said.

Hesiod, poor, despised, lonely, felt the wooing of some higher influence and called it Spirit, or Breath of Heaven, and he founded a new school of poetry with something like the power of "thrice ten thousand spirits round our pathway gliding, rewarding some with glory, some with gold," as the key teaching of his successful school.

Joshua set up twelve stones in the midst of the Jordan River to commemorate the dividing of the waters by heavenly influence where the feet of the priests bearing the ark of the covenant of the Lord of all the earth stood firm, rewarding the Hebrews with what Hesiod's poetry would call glory.

The miracle-working Unseen had been the glad theme of inspired men before Joshua and Hesiod. They all taught that a great unseen world swings close to this world, an imperishable world, whose substance is Zeus, or Brahma, or Ra, or God, or Gaea who sprang from Chaos, or Cimah the queen of heaven to whom the Hebrew women offered cakes and worshipping incense in the streets of Jerusalem. Jeremiah was angry with the women for not calling the ruling other world that swings so close to this world, *Jehovah;* but the women said that they had had plenty of victuals, and had been well, and had seen no evil while worshipping the queen of heaven,

the nourishing Cimah, and they would not call the miracle-working Highest One *Jehovah.*

The breath of the unseen world was in the refresh-
ing winds that hailed from above the primordial men
as they bared their heads and purposely inbreathed the
ether winds subtler than earth's common atmospheres
which to this day wait actionless for man's inbreathing
action:

"Breathe toward me heavenly breath, till all this
breathing form of me quickens with breath divine."

Father John of Russia felt the stirring of a breath full
of curative elixirs quickening the bread and wine of the
eucharist. He cited cases of bodily healing while par-
taking of these ether-charged elements. Certain names
of old were reputed to liberate the healing elixirs biding
forever within the airs we daily breathe never noting
their mystic offers: "Why, O man will ye die, having
power to partake of the breath of immortality?"

Even watchers toward the winds that make forecasts
for coastwise shippings and aerial flyings declare that
there is one wind that comes from above which always
fetches reviving. It is the northwest wind. But to them
that seek all their good from above every wind contains
over-breaths of willing revivings or buoyant ethers of
renewal. Ezekiel called for all the four winds to breathe
forth their secret vitalizings that the multitude of dead
in the Seir valley might quicken through all their earth-
bound forms.

The cradle doctrine of the New Age is, "Thou God
seest me—For I also have looked after Him that seeth
me," as it was the cradle doctrine of earliest Hebrew
prophetics, announcing the coming together of aristocracy and serfdom in the Christ above the pairs of oppo-

sites rich and poor, bond and free. Every man's hand has
been against Ishmael the bond child, and Ishmael's hand
has been against every man, yet Ishmaels' cradle instruction was "Thou God seest me," and the Ishmael type did
found a great nation and place twelve princes in the earth
whose determinations are with us to this day.

And Isaac, the aristocracy of the world, with the same
father Abraham that the bond child owned, though he
has held the world in fee, must with the serf begin again,
"Thou God seest me—For I also have looked after Him
that seeth me," that the time of their warfare may be
shortened according to Christian decree.

Abraham their father looked for a city which hath
foundations, and sought the wooing Unseen by looking
up and hearing the voice of the Light of the city saying,
"The land thou seest, to thee will I give it."

Why should serfdom and aristocracy fight for land,
or houses, or precedence, when all that belongs to either
of them they may inherit by lifting up their eyes to the
world that swings close, whose miracle-working winds
may any time waft in miraculous new prosperings?

"The spider taketh hold with her hands, and is in
kings' palaces," says Solomon, speaking for those who
seek their highest *best* by highest *watch*:

"By so many roots as the marsh grass sends in the sod,
  I will heartily lay me ahold on the greatness of God" *The
  Miracle-working God.*

Mystical Science brings the close swinging miracle-
working uplands to our attention with perpetual urgings.
To bad and good among us it declares the same: "Lift
up your eyes to the fields white for the harvest." If we
keep off the world's fighting ground of good and evil in

thought and conduct, the clinches of its estimates of what is good have no yea in us, and its estimates of what is evil find in us no agreement. Visioning toward the same country Abraham was seeking we swing in with the Miracle-Working Heights. "The land which thou seest, to thee will I give it," whispers its unceasing voice. "Honor and fortune exist for him who remembers that he is in the neighborhood of the Great," wrote the truth-announcing Emerson, keeping step in a mystical moment with Moses the God-taught law giver, and Ezra the inspired reformer.

Millions perished marching toward the Holy Sepulchre to take possession of it. Where was their inward watch? Remember that "We always look toward an object before thinking it, and it is by having oft recourse by inward viewing that then the mind goes on to know and comprehend"—and the body, obedient follower, to show the concrete thereof. Who on earth was there to teach the pilgrims and crusaders, sepulchre-ward inward viewing, that there was another country smiling over their heads, whose giving was prospering, unsepulchred, joyous life?

He of highest vision rises highest even among the sons of earth. Samuel said, "I am the seer," and he went forth to choose a king for Judah and Israel. Jesse of Bethlehem-Judah was the greatest man among the Hebrews, and his sons the finest specimens of Judean manhood. Eliab, Abinadab, and Shammah were marched before the seer. One had a genius for war, one a genius for statecraft, one for language. What a wonderful gift is language! If your language were right you would be the desideratum of kings' courts. You would stop the warfare between Isaac and Ishmael, capital and labour, rich and poor!

Samuel Johnson, uncouth, untidy, ragged, held the young peers of Oxford spellbound with his speech, and the masters gave him larger freedom of action than his confréres, because his words cut swaths like polished scythes. Djemal Pasha ruled the Turkish empire with his speech. The world awaits its master of brotherhood-cohering words. "A right word how good it is, who can measure the force of a right word?"

But the Lord said to Samuel, No, not one of these for king of Judah and Israel. For the one with genius for war has his inner vision toward men to marshall them for martial triumphs; the one with genius for statecraft has his inner viewing set toward men to place them as heads of tribes and territories, and the one with genius for compelling speech is observant of men's swaying emotions under his spellbinding oratory.

Now their youngest brother David took not much note of men. "Mine eyes are ever toward the Lord," he said, "for he shall pluck my feet out of the net." Not what he himself did, but what the Lord on high was doing, was David's aspiration. And the seer anointed David with kingship; and he was the greatest warrior on earth, the greatest statesman, the most victory-awakening speaker: —"The Lord hear thee in the day of trouble; the name of the God of Jacob defend thee; send thee help from the sanctuary; and strengthen thee out of Zion," was his national rallying cry.

In the time of the World War one nation shouted, "We are ready to sacrifice seventy-five thousand men any day." And it was soon recorded that that nation had sacrificed more men in proportion to those sent forth to battle front than any other nation engaged in the great conflict.

What should that nation have shouted? "We are ready any day to forward an army to the triumph of our righteous cause!" Why should they have called the attention of our globe to sacrifice? Are we not soon showing forth that to which our inner gaze is directed? Sepulchre 1000 A. D. or Sacrifice 1917 A. D.?—Who shall rouse the world to choose the sepulchre-and-sacrifice-undoing *high watch?*

Nothing shall by any means defeat us lifting up our eyes ofttime to Thee!

Elisha was enraged at Joash king of Israel, for not smiting the earth with his arrow five or six times, instead of only three times, when Joash was told by Elisha to smite the earth with his arrow. For *five* stands for successful labour, and *six* stands for labourless success, or the miracle of The Highest ever moving hitherward. The hyacinth bulb planted top side down labours desperately to bloom Chinaward, and succeeds in making an anaemic blossom, pale and small, but plain imitation of itself as it would be facing the sunshine and ambrosial morning airs. What does the downward looking hyacinth know of its own sun-facing, unanaemic wide-spreading beauty? So we have never seen a man of the Labourless Supernal order, facing the "Sun of Righteousness with healing in his wings," till the New Dispensation gave us its good news when he came toward us! An inspired full victory, or the labourless miracle, was Elisha's hope for Joash. But Joash had only three small victories over Ben-Hadad king of Syria, enemy of Jehovah, as his three arrow-smitings pre-figured—wide indeed of the mark of the high success his royal right might claim.

"I will fetch my knowledge from afar," said Elihu, youngest of Job's advisers. "Behold, God exalteth by his power, who teacheth like him?—He scaleth instruction while man slumbereth on his bed—He openeth the ears in the night time—Remember to magnify his work, that men may behold it afar off."

The more we look to Ain Soph the Great Countenance of the Absolute, the Origin of knowing, the more we know of the influence of the close-swinging miracle-working world in its finished beneficence:

"Listen, O isles, unto me; and hearken ye people, from far"—"I will show thee great and mighty things, which thou knowest not."

Solomon got him to the myrrh mountains of wisdom for the mystical daybreak, and to the hills of frankincense for the streams of healing hailing from above the sunrise. Beyond the margins of the mind supernal Wisdom waits. Over the Dayspring of a glory from above, the heavens' mysterious healings are distilling for those who obey the mandate—"Lift up your eyes—Look unto Me—I am the Lord that healeth thee."

The greatest doctrine announced by the voice of inspiration, is that a Mighty One onlooketh us, wooing ever as, "What wilt thou?"—"Ask what ye will"—"Is anything too hard for Me?" Who of us, uplooking, answers face to face as a man talketh with his brother? Who of us holds high converse with the Ruler in the heavens, who saith, "To him that ordereth his conversation aright will I show salvation?"

In ancient mythology the sixth god was Cronus, god of the harvest, with two faces, one glowing white and one glooming dark, according as he set himself to harvest from the heights above or from the depths below. When the giants got six-toed from earth bound attentions they were destroyed. When the Man of Galilee touched the six-stepped throne place by daily converse with what

David called *The glory above the heavens,* such magian effulgence gleamed forth from him that multitudes coming unto him, he healed them every one.

We are all harvesting according to our inward viewings. "Why are ye troubled?" and "Why do thoughts arise in your hearts?" asked Jesus. We had to wait for the birth of Hegel, 1770 A. D., before we got the scientific answer to that question: "We always look toward an object before thinking it," wrote Hegel, "and it is by having oft recourse by inward viewing that then the mind goes on to know and comprehend."

High mysticism is not a science of right thinking, or right conduct; these are strenuous labours. High mysticism is the call to look up to what the Cabala of Jewry named *Ain Soph The Great Countenance of the Absolute,* who ordereth thoughts and speech and conduct anew. "Behold, I make all things new," He saith. "I will return unto that people a new language." "They shall speak with new tongues."

The speech of man always declares where his inward viewing, laden with harvestings for life conditions, oftenest alights. The good woman mentions the shiftlessness, the sorrows, the sufferings of her neighbours, and she has some unpleasantness of body and of affairs formulated by the inward gazing her verbal descriptions betray. Is it not written that the wages of sin (or aberrated vision) is death? Mysticism is not a science of goodness and badness of conduct. It is the science of that which harvests as good or bad conduct. It is the science of the genesis of conduct and the genesis of thoughts. Secret viewings compose their own speech, rouse their own emotions and formulate their own actions.

Our initial and compelling faculty is our inner vision.

Vision often Godward and live anew. So shall the body
"be like a tree planted by the rivers of water"—whose
leaf fadeth not. Vision often Godward so that affairs also
may go well. Gaze often toward Our Father, and all
thoughts shall be like morning music. Lift up an inward looking now and then to a country whose ether
winds ever raying forth their healing aura, are fleet
remedials for all the world's unhappiness.

Does not Dr. Jowett record that Bishop Westcott of
Durham was a man of royal strength from gazing toward
the glory of the Highest? How did he know there was
a Highest to gaze toward? How did he know there was
any glory to harvest from as royal strength of mind and
body? What was he gazing with?

The kings of Tyre wore the sardius stone, sixth stone
of Revelation, to signify that they had caught a threefold
lustre by invocation. The sixth stone of Revelation is
the lustre sardius, symbol of invocation.

By physical invocation the body may renew with the
subtle elixirs that wait to mill within it to strengthen
and uplift. The hills open in the daytime and at night
pinch into themselves the circumambient vitalizings with
which the daytime airs are charged. Then they bring
forth grass and herbs and grains for cattle. The brain of
man can open and close into itself the wisdom elixirs that
hug it down like Sibyls' hoods. "Lift up your heads, O ye
gates . . . and the King of glory shall come in. Who *is*
the King of glory? The Lord strong and mighty . . . The
Lord of hosts, he *is* the King of glory." There is no part
of the body that may not open and indraw the subtler-
than-air stimulants that would soon burst forth as
strength. This is physical invocation.

The mind is surrounded by distant knowledges that

it may indraw by asking of awaiting knowings, and burst forth with answers as from unseen teachers. This is mental invocation. The sardius stone stands for mental invocation.

The lustre sardius stands for mystical invocation—the way to know new things from the heights above the margins of the mind, where things hitherto untaught lie waiting our mystical invocation. Pythagoras said that there are magic-mighty syllables making up an Ineffable Name, Key to the mysteries of the universe. Speak them as into upper ethers where the hitherto unknowns abide, and they will mill hitherward with new knowledges. "The Ineffable Name is Key to the mysteries of the universe."

The Parsees called the magic syllables so equipped with drawing power, *The Ardai Viraf Name, full of Revealings.*

"And the throne had six steps." "And I saw in the right hand of him that sat on the throne, a book."

Every man's name is the book of his equipment, and the use he has made of his name constitutes the influence of his name, or its spirit, or its ghost. Would we not expect that he whose vision had lifted highest, and whose invocations had brought him to highest harvestings would shed the strongest and best influence, or spirit, or ghost, by the calling of his name? He who sat upon the throne, with the book in his right hand, announced to all the world, "The Holy Ghost, whom the Father will send in my name, he shall teach you all things."

Surely the magic name for which Pythagoras was seeking to reveal to him the secrets of the universe must be concealed in the name promised to reveal all things by invocation! From its breathing forth power the Parsees might have taught the wisdom-wafts their longing aspirations forecast.

"I wept much," said John the Revelator, because the world went on so long without noticing that some names are like alabaster boxes of empowering ointment. Break them open by invoking them, and mysterious new influences awake through body and mind and environments.

Paul wrote to the Ephesians that revelations would be vouchsafed by the acknowledgment of Jesus Christ, and the eyes of man's understanding should open to know the hope they may entertain from the calling of him that hath put all things under his feet, with name above every name that is named, not only in this world, but also in that which is to come.

The one insistence of the Jesus who was Christ with the fullness of the Godhead bodily, was that His name sheds forth fresh life: "My words (my syllables) are life." "I came that ye might have life." "I am the living bread." "He that eateth me, even he shall live by me."

Tennyson wrote that "'Tis life, more life for which we pant," but he did not tell us how to get more life. Only the Christ Jesus man ever told mankind how to quicken anew with the "life of which our nerves are scant," and He declared that His name would stir the ether-breaths of the unseen eternal world to inbreathe with our inbreathing, as the ancients called the names of mighty warriors and wise saints and drew toward themselves their influential vigours.

There is no science of wealth or health so great as the science of inbreath, was the conclusion of the ancients.

"An hundred forty and four thousand, having His Father's name written in their foreheads," said John, which means that the right number to stir the world shall have cognizance of awakening Spirit, or Holy Ghost, or healing winds, or God breath from heaven that the name of the Lord of Life conveys.

> O Hither wafting breath of strength,
> There's a heaven-taught way of reaping!

Shall we know less about invocation than the Greeks and Romans of old, calling, "O Mercury, grant me magistral to poverty—O Morpheus, come with celestial soporifics!"

Invocation is the greatest efficiency of prayer. Did not the voice of far past inspiration declare that the name of Deity transcends prayer? Jesus of Nazareth called the Overlooking Deity *Father!* At the crucifixion He cried, "Father, forgive them!"

Why does the world persist in telling itself that the cry of the Jews invoking the blood of the cross to take vengeance on themselves and their children has been the only prayer answered through all these centuries? Why have we not been taught to lift up our eyes to the prayer-answering, for-giving FATHER, and partake of the now hitherward streaming Beneficence into which the glorified Jesus was then gazing? Should the offspring of the mistaken Crusaders be harassed world without end? Should our children's children be taunted on and on because of our today's foolhardiness? Though your errors were as scarlet, the for-giving answers to the Jesus Christ prayer on Calvary passing them along to the sin-undoing High Redeemer, they shall be as if they had never been.

"I will set no wicked thing before mine eyes; . . . it shall not cleave to me," was David's song for our instruction.

In the science of numbers, *Six* is significant of complete manifestation, all the faculties raised to their highest exercise. And Sin is errant vision unified with errant words and acts. Errant vision, or sin, carries the banner of triumphant arrival at complete manifestation.

To the mystically wise the badge of triumphant arrival at the worst that downward vision can accomplish, is tantamount to telling plainly the possibilities of upward visioning.

To the mystic the victorious arrival of a bodily swelling, the result of somebody's downward viewing, at the strength of its ultimatum bursting and blasting with death, show how sight upward toward the Author of strength would make manifest Undefeatable Omnipotence. "For the way of life," wrote Solomon, "is above to the wise, that he may depart from hell beneath."

The secret that sin tells to the wise, is that by vision toward matter and mind and their laws of pleasure and pain we get caught in the wheel of destruction, but by vision toward the Highest One above the pairs of opposites, we get charged with independence of matter and mind.

To the mystic, the sensualist with his blazing sores of malefic contagion, is sign royal of the possibility of one charging himself to fadeless bloom with heavenly inspiration, till men round the globe to touch the hem of his garment of healing; till "the gentiles come to his light, and kings to the brightness of his rising."

There is but one commandment issuing from the Giver of Almighty Health, and that one commandment, is "Look unto Me, and be ye saved, all the ends of the earth."—"Beside me there is no Saviour."

Where did Elijah get strength to run barefooted all

the twelve miles from Mount Carmel to Jezreel, ahead of King Ahab's swift horses, leading them by bridle and tiring not? From inbreathing winds of unceasing strength from above, where his inner eye was seeking the Author of strength, who giveth to all men liberally, even to unbreakable Omnipotence. "As the Lord God of hosts liveth, before whom I stand," he said.

Elijah did not touch the *six* mark of arrival at the best that the high watch could do for him, for he did his inbreathing of victorious energy intermittently, and unwittingly, and his upward watching halfway, unknowing of the secret of uplooking power. Like most human beings of this very day he spent more time viewing his troubles and describing them, than glorying in the immune splendour of the wisdom-imparting Jehovah.

To the true mystic the upward looking and the inbreath of heavenly atmospheres are volitional and scientific. He knows whence the energizing breaths come wafting their newness, and the healing elixirs come streaming with enlivening vigours, and he knows that if he is not rounded up with unkillable Omnipotence it is because he has neglected his science.

The world waits the *six* mark of unceasing acceptance of the divine Highest only; it waits the visible bloom of immortality in the Garden of Man—the Jesus point of plain demonstration that "No man taketh my life from me."

In the Apocrypha, we read that in the sixth place the Lord imparteth understanding. Understanding imparted from the Lord is the desideratum of man, for having touched this magian flame his life is kindled to glow forever. For "understanding is a wellspring of life." Mystic understanding is strength, identical with divine strength. "I am understanding," saith the Lord, "I have strength"; and it is written that He giveth this strength to His people.

Habakkuk, reputed to be the Shunamite child raised from the dead by Elisha, could not help reasoning, in his first chapter of prophecy, that because the Lord his God was from everlasting, *he* need not die, though the Chaldeans, supping up their leopard-like strength from the east wind, scoffing at kings, and deriding every stronghold, should become dust heaps like as their swift horses.

In self-forgetting moments Habakkuk breathed from beyond the margins of the Chaldeans' powerful east winds. He breathed the ethers that lave the throne steps of divine arrival. He made it education to know sin's glorious secrets as his three mystic chapters reveal them, telling of the high watch and the low watch, and of saving and gathering and honouring as the time of unspoilable wholeness ripens, disclosing the bloom of arrival at the glory of the Lord's still, secret inbreaths, mighty to save.

Habakkuk's three chapters lay stress on the end of captivity, everywhere visible to the eyes of all watchers toward Him that saith, "Look unto Me." They tell the story of the Holy First, who came from Paran, the hidden heights, whom the Gadarenes of earth implored to depart again into hiding. They tell how, when sin's vast secrets uncover, mankind shall choose to breathe from above, till they all bloom as unspoilable health made visible, in comradeship with the Lord of fulfillment, and the brightness of His completeness shall not be lost to their view.

These three chapters of Habakkuk tell of the presence of the Lord of demonstration now among us, back of the coasts of Gadara, or our unopened faculties, ready now to quicken to life in every part all who believe on His leadership and His nearness, and call upon His revealing Name. "Who hath believed our report?" (see Fifth Study) "That believing, ye might have life through His name." "Call unto me," He saith, "and I will answer thee, and show thee great and mighty things which thou knowest not—and all the nations shall fear and tremble for all the goodness and for all the prosperity that I procure unto it."

Paul was told to rise up quickly, calling upon this Name, and he knew its ripening value, as the Gnostics of his time knew the gift-giving name *Abraxas*. "I press toward the mark, for the prize of the high calling of God in Christ Jesus," he said, "and if in anything ye be otherwise minded, God shall reveal even this unto you."

Many Christians have been "otherwise minded," in their calling, and have had what they called for revealed in full measure. Carlyle called, "O Fortune! Thou that givest to each his portion on this dirty planet! Grant me literary distinction!" And he declared that ever since he had been able to frame a wish, the wish of being famous had been foremost.

Eliphaz the Temanite enjoined upon all mankind the practice of the calling principle: "Thou shalt lift up thy face unto God," he said, "Thou shalt also decree a thing, and it shall be established unto thee." Every choice is a call, direct or implied. Carlyle's call was so direct that literary fame was promptly "established unto him." William the Conqueror's call was not direct like Carlyle's, but so strongly implied that he laid hold of England with uncanny power, defeating the brave Saxons with astonishing ease.

He had for years been claiming the crown of England as the bequest of Edward the Confessor to his father. Landing at Pevensey near Hastings, with 60,000 warriors, to take possession of his vision's crown, he fell forward upon his face into the sand. Then all his 60,000 attendants blanched in their faces, for to be apparently *hors de combat*, biting the dust on a strange coast, was evil omen. But William's vision was too strongly set toward England as his own possession to regard omens. Clutching the sand in his fingers, he used the words that tallied with his steadfast view: "By the splendour of God, I hold the soil of England in my hands!" he shouted. The historic demonstrations of important men give definite cue and clue to each individual's demonstration, on small or large scale, according as his choice has been powerful or puny, and his objective glorious or commonplace.

Paul and the Christians he gathered round him lifted up their faces to the Almighty, as Eliphaz had enjoined, and called to the Victorious One, till they were great conquerors together. "I am more than conqueror," Paul declared of himself. And he insisted that the victorious Christ Jesus he had so persistently invoked, charged him so near to the brim with Christ executiveness, that he could work miracles. "I can do all things through Christ which strengtheneth me," he said.

Paul did not round up to the *six* mark of unkillable Life and unbreakable Omnipotence, because of getting entangled in a downward watch toward sex differentiations, and foods suitable and unsuitable. He split on the rocks of sex and food. Though on his spiritual flights he proclaimed "neither male nor female . . . in Christ Jesus," yet he suffered not a woman to speak in the Christian Church. Though on his vision's scientific formulas he read plainly that, "For neither, if we eat, are we the better; neither if we eat not, are we the worse," yet he became explicit in explaining what might and what might not be eaten, and forbade any man who was not a strenuous worker to eat anything at all; as though eating were very important.

The glory of sin is its consistency—it keeps at its own stride toward full arrival and nothing diverts it from its *six* mark's grand completeness. Does any upward watcher catch heavenly health by his persistence heavenward, till his health breaks forth like the morning, and every sick person catches health from him, as surely as the downward watcher identifying with small pox, gives his neighbours the same, if they come into contact with him? It is only the Jesus among us of whom is it yet written, "Multitudes came unto him, and he healed them every one." "But as many as received him, to them gave he power to (visibly) become the Sons of God."

In mysticism we learn the law of our subtle visional sense. We find that we use our inner vision constantly, and we find that our thoughts follow its wake, and that physical conditions follow the thoughts. By knowing the law of the inward vision we follow intelligently Job's lamentation that his thoughts were only unmanageable shadows, till his witness was in heaven, and his record plain on high. We know how the seers read right descriptions of all that is transpiring far or near: "It was the labour of mine eyes," wrote Asaph the seer, after studying into the secret workings of his successful seership.

For whatever is looked toward as an objective to the

inner eye reveals its secrets, whether the Lofty One inhabiting Eternity, or the pyramids of Egypt, or the
motives of our neighbours. There is but one law running
along to the flowering of all steadfast visioning.

"The secret of the Lord is with them that fear him."
"Fear" is singleness of eye. "In the fear of the Lord is the
instruction of wisdom." "The fear of the Lord is a fountain of life." "There is no want to them that fear him."
Fear, or singleness of eye toward any objective, is dis-
closure of its secrets, and experience of its working power.

"Pass the time of your sojourning here in fear," wrote
Peter, after having wrought great miracles by obedience
to the Mosaic law of fearing the glorious name of the
Lord. Peter discovered with the prophet Micah, that
there is great effect from singleness of eye toward names,
as toward other objectives, and he chose the name of the
Saviour of men to stand by. Micah got it as an unfailing
law, that, "My name shall (cause to) see that which is."
And Peter proved it by walking with the angels, and
doing their works, after fearing the "only name given
under heaven, whereby men must be saved." For as the
angels do wondrously, so also did Peter wondrously always by the power of the Name he inbreathed.

The purpose of religion has always been to teach
men to make the most of themselves. And at its acme of
instruction it has taught that right thought and right
conduct follow right view. To look steadfastly toward
the Helper, and Healer, and Saviour abiding unseen but
always within calling distance, has ever been its injunction: endure, "as seeing Him who is invisible."

The Gnostics of old endured seeing and calling toward
Abraxas, an unseen giver of tangible blessings.

The name of one who has accomplished great works vibrates with his genius. Aristides, the wise archon of Athens, knowing this, set his eye toward Æsculapius, whose skill in curing diseases and restoring the dead to life, was traditioned to have made Zeus angry, lest he rendered men immortal; and Aristides called faithfully upon Æsculapius, till he came swiftly through the formless spaces, and stood in his presence, audibly counselling Aristides for the successful cure of his bodily disease.

"There is one operative virtue running through all things," was a discovery of Cornelius Agrippa of Cologne. "All things supercelestial may be drawn into the celestial, and all things supernatural may be drawn into the natural," was also his discovery, though the practice of drawing down the virtues ascribed to the different angels, into statues, and periapts and wafers, long antedated Agrippa's explanation of the magical drafts men may make on the good offices of nature and of God.

By much attention to material things, men make such drafts on matter that its mysterious operations seize upon them in more than conceivable manifoldness. They suck in the way of matter, by indirect calling, as they look to it for their welfare and their knowledge. This is the way of which Solomon wrote, "that seemeth right unto a man, but the end thereof are the ways of death."

Let men but volitionally lift up their eyes to the Sender of the mystic cure currents ever hailing hitherward, and make all their drafts on these immortal streams, and the mysteries of health, and prosperity, and incessant renewal shall be revealed and experienced. Jesus having demonstrated this, his name as the vanquisher of death, and the abode of all gifts of God, made him able to declare, "Whatsoever ye shall ask in my name that will I do." "Ask what ye will, and it shall be done unto you." For "all things are delivered unto me of my Father."

"This is the path which the lions' whelps (or the obedient followers of the masters of material science) have not trodden, nor the fierce lions (the masterful Kings, and Generals, and Presidents among us) have not passed by." It is that lost old superstition, in which the world was nearer essential truth than ever in its so-called scholarly acquirements.

The Science of God reveals all science. The true Name of God reveals the Science of God, and reveals all names, from the names of the stars to the names of the insects, from the names of the fearless archangels to the right names for the victorious earth walk of our children.

The name of one who knows the Great Revealing Name, reveals his knowledge. "The Holy Ghost, whom the Father will send in my name, he shall teach you all things." "Thou holdest fast my name—I will give a new name which no man knoweth save he that receiveth it."

All that we as yet know of the Maker of the universe is the practice of His Presence. Of His actual substance and purposes we know nothing. "Touching the Almighty we cannot find him out—with him is terrible majesty," reported Elihu, who yet found the Almighty teaching and exalting him, and giving him songs in the night, as the result of his upward watch and the inbreath of the Almighty that waked the God-Seed of understanding within him.

Elihu seemed to know no other name of the Highest but "God." But he spoke of a messenger, an interpreter, a ransom, an atonement, one who could enlighten with living light. He could not speak the name, but fetched, he said, the knowledge from afar, that by the recognition of the atonement, man should begin to breathe in lakes of joy and wisdom from the Giver of wisdom, thereby comrading on terms of equal arrival with the Man whom the race of men had accepted as the flower of divine arrival on this globe. The recognition of the Man of atonement, makes draft on the Holy Spirit, "The Holy Spirit whom the Father will send in my name." It makes draft on life: "I will put my spirit in you and ye shall live." Wherefore turn yourselves, and live ye.

The Holy Spirit is the ghost or spirit of wholeness; or breath of wholeness. Job speaks of the spirit of God being in his nostrils; and Job's whole life was changed by refusing to speak against the transformation which the inbreath of the Healing Ghost was making in him. Job was faithful to the high vision and the atonement, though he knew not the name of the Daysman who should make the atonement. Evidently Job knew that to regard the breath he breathed as the Healing Spirit, instead of common atmospheric air, and never to speak of the outcome as anything but transforming, would fetch him out on to some upland that men who breathed just air, and called it air, would never reach. And he was right. His witness in the heavens and his breath always known by him while inbreathing it as the God breath, brought him to where he was blessed beyond his calculations, and with mysterious wisdom he knew himself well pleasing to the High Eternal.

The Apostles of the Regeneration knew the name of the Daysman, and they knew their breath as charged with the ether wafts of healing, and they knew the law of the vision, and the reborn status that was to be the outcome.

This was their Christianity. In the name of the Great Achiever they drew their breath as the Holy Spirit, which is the spirit of wholeness, and they were transformed characters. They walked in the fourth dimension in space, and no prison walls could hold them, and no lions' jaws could destroy them; and no former ignorance or common birth among the people who breathed common air, and sought their life from material productions could count against their being the wisest men on earth. They took up a life near us and yet above us, a life with the power of manifestation here and there, and now and then, throughout the generations, not unmet even in this twentieth century. But their mode of arrival at the state of just men made perfect, on the *unseen plane,* is not the final arrival that mystical interpretation foretells.

To be charged to overflow with irresistible miracle-working while yet manifest in the flesh, to be the radiance of buoyant joy while yet *walking among the sons of men,* to shed the perfume of healing and strengthening and illuminating while yet *speaking with us and smiling upon us*—this is the final Christian ministry; this is the bloom of full obedience to the Sacred Edict, "Look unto Me."

There is no warfare where the vision of God is. There is no disease where the healing Name is called. There is no inadequacy or failure while the spirit of God is in the nostrils, inbreathed as the only breath. This is living truth.

If Pope Sixtus, as a ragged lad under the trees, could lift up his face toward the papal throne, and decree to be pope of Rome, and draw himself by his decree to the papal seat, equipped to perform all its high offices with intelligent determination and successful energy, so some lover of mankind can lift up his face to the Almighty, as Eliphaz counselled, and decree to be the healing and strengthening and illuminating of all the people whom he may look upon, or who look upon him.

If we would transcend our limitations, we must look above our limitations. And we always believe in what we look toward; and we draw what we believe in. Deep believers in punishment are unconsciously drawing it. All the prophets drew punishments.

Great believers in miracles draw them. George Fox, Evangelist Finney, and Dwight L. Moody, drew pentecostal tongues, and spoke languages they had never studied. Irenaeus bishop of Lyons wrote that in his time there were "many brethren speaking all kinds of languages by the Holy Spirit." And St. Sauveur of Horta cured six thousand persons during the feast of the Annunciation, through his devotion to the healing grace his inner eye beheld and his mind so ardently decreed.

The watch of the first Christian Apostles was toward the Lamb that in the midst of the throne is man's unfailing Provider; and they wanted for nothing, and all those who gathered to them were abundantly supplied. Their vision was toward the Highest, and no man could set upon them to hurt them, with power to defeat them, nor hurt the thousands obeying their injunction to continue in prayer and watch in the same. Their faces were set toward the Author of peace, and they allowed no man to be a slayer of himself or his fellow-men. Their look was toward Christ the Triumphant, and no man could grovel in their presence, no man could lack joy, no man was foolish. They were the spiritual magistral, or sovereign remedy, answering the prayers of the ancients for a way of universal cure, opiate divine to the sorrows of the world.

The High and Lofty One inhabiting Eternity is above the pairs of opposites, good and evil, life and death, spirit and matter, therefore his ways that inspire us as we look toward Him are not our accustomed ways. We let Spirit go on its free way, to leave us poor in spirit; we are independent of Good; we do not hug tightly to Life, as though it were precious. For Spirit, Good, and Life, are gifts of the Highest.

Shall we be in love with gifts, like Nebuchadnezzar at the gate of the fiery furnace, calling for Hananiah, Mishael, and Azariah, whose names stand for the gifts of the Highest, instead of the Highest? It is not the gifts of the Highest that say, "Look unto Me and be ye saved," though all the gifts of the Highest do come if we set our inner eye toward them and call their names, drafting on their waiting offers. Can we not call sleep, and inbreathe sleep, that subtle thing that Ezra said the beloved of God receive, till insomnia flees our being, and the poppy breath of heavenly forgetfulness wafts us into Elysian fields? But sleep is not God. Sleep is a gift of God. To joy in sleep is to be like Isaiah's people joying in a harvest of grass and apples. The bloom of sleep is not the perfume of unspoilable health shedding itself abroad from the sleeper. We can call strength, inbreathing its offered vigours, looking toward its omnipresent smile, till strength nerves us to masterful handling of lions and elephants. But strength by itself is not God. Strength is a gift of God. No beams of the Almighty radiate unceasing renewal to the fainting from the heaviest weights among the powerful pugilists.

To joy in strength is to joy as in a harvest of apples that lose their flavor, or corn that moulds. For all the gifts, sought with eagerness for their own sake, have their round of existence and disappearance, and those who receive them partake of the same. But thou! O Highest Original! art forever, and the manna Thou givest for the calling of Thy Name, is radiant beneficence scattering and yet increasing, till the world is alive forevermore!

The choice of the High Original alone instructs the mind, renews the body, engirds the affairs. "Call unto me, and I will answer thee,"—"revive my Spirit in thee," and—"nothing shall prevail against thee."

Nothing prevailed against Jesus. "I can both lay down my life and take it up," He said. And He knew all things, and needed not that any man should teach Him. His face was always heavenward, comrading always with the King of Kings and Lord of Lords. Thus invoking the name of Him calls toward us His masterfulness of life and death, knowledge and ignorance, health and sickness, majesty and insignificance. Has it not always been believed that invoking the name of a man of masterfulness and courage inbreathes masterfulness and courage?—

"Go, my dread lord, to your great grandsire's tomb,
  Invoke his warlike spirit." "Let the king hear us when we call."

And that calling sleep brings it?

"Draw near and touch me, leaning out of space,
      O happy sleep!
  Enfold me in thy mystical embrace,
  Thou sovereign gift of God, most sweet, most blest,
      O happy sleep!"

And that calling the invisible goods with which the spaces are charged fetches them?—

"I will call for the corn and increase it."

"I have made thee like unto him even God, who . . . calleth those things which be not as though they were."

Call to the trees that lean and whisper against the far horizons, and they shall tell thee all their secrets, from the cedars of Lebanon to the hyssop that springeth on the wall. Call to the stars that lie on their black beds through the long north nights, and they shall tell thee what the schools have not discovered.

Call to the Lord of Life and Glory beyond the bars of human sense, and all the living fountains lying deep in thee shall quicken into stirring streams, and all the pent-up wisdoms that inhabit thee shall leap to meet the Universal Wisdom, shining forth as the sun with the glory of their Father.

"There is none like unto thee, O Lord! Thou art great, and thy name is great in might!"

When the disciples had associated with the Risen Christ till they knew that He had borne the griefs and carried the sorrows of all the human race, and by reason of His God Substance it had been as nothing to Him, then "opened he their understanding, that they might understand the Scriptures." He dissolved the bars of human ignorance that hid their Lord-implanted inward wisdom. He dissolved cold foolishness and dark ignorance by the hot beams of His bright Righteousness, or shining Understanding. Did not Aristides say that when he associated with Socrates he felt himself flashing with wisdom? Did not the Chaldeans say that when Daniel came near them, their doubts were dissolved? Does not the fearlessness of a bold and successful leader communicate itself to his associates?

The Risen Christ was weightless of body and thinkless of mind. He was Pure Understanding, hot with universal dissolvent to human heaviness, dewy with working mystery. Then was fulfilled in His disciples the inspired promise that they of understanding should do exploits. Then was fulfilled in them the promise that whosoever should draw out his soul to the hungry, and satisfy the afflicted, should have his light shine forth in obscurity, and the Lord should guide him continually. "To satisfy the afflicted" is to acknowledge the Lord of affliction. The Lord of affliction was Jesus the Judean, wounded for the transgressions of the world, bruised for its iniquities, bearing the chastisement of the peace of all mankind, that by His stripes they might be healed.

To draw out the soul, and to acknowledge, are noted as identical activities in the Scriptures, being followed by the same results—the breaking forth of the light, and the consciousness of daily guidance by the Lord of victorious living. "In all thy ways acknowledge Him, and He shall direct thy paths." "Draw out thy soul to the hungry, and satisfy the afflicted . . . then shall thy light rise in obscurity, and thy darkness be as the noonday: And the Lord shall guide thee continually."

To acknowledge the Great Hungry is a mystical expression, meaning to acknowledge Him that swalloweth up death, and hell, and the "Egyptians," and the ways of the paths downward. This is to acknowledge Jesus the Saviour from death, and hades, and darkness, and the paths downward.

And the only Scripture the Risen One gave the

disciples, after their acknowledgment of His vicarious achievement, was His own Name. "The Holy Spirit whom the Father will send in my name shall teach you all things." And to the world at large he prophesied: "Ye shall not see me henceforth, till ye shall say, Blessed is he that cometh in the name of the Lord."

No library on earth holds a book guaranteeing to teach all things. But the ancient wise men declared that there is an Ineffable Name that teaches the mysteries of the universe. The Rabbis said that Jesus of Nazareth did not know the Ineffable Name, therefore His miracles were wrought by sorcery learned in Egypt, not by the power of the Great Name. Matathia, in the Nizzachon says this, writing for the Rabbis. But Jesus himself testified for all time, "The Holy Spirit whom the Father will send in my name shall teach you all things." "In my name preach the gospel, heal the sick, cast out devils, raise the dead."

He knew the Ineffable Name that is key to the mysteries of the universe. And He knew that whoever should keep His Name as Jesus Christ should come into the Ineffable Name.

"Thou holdest fast my name—I will give a new name." This new name cannot be spoken without instantly accomplishing the raising of the dead or the healing of the sick, or the illuminating of the life. It can never be spoken in vain. "Thou shalt not take the name of the Lord thy God in vain," is a prophecy, like "They shall not hurt or kill in all my holy mountain."

All the so-called commandments are prophetic utterances, to the Initiated. They all mean that when the Ineffable Name of the Lord is known, we are at home in the perfect land, where the former things come not into mind any more.

The Valentinians by a cabalistic system, *notarikon,* made the name *Jesus* the equivalent of Jehovah Shammain, or the saving Word; and Osiander the Lutheran studied the two syllables as in themselves the Ineffable Name.

That the name *Jesus Christ* is even at this day, and in this age of regarding the historic sufferer rather than the victorious Peace-Presence, a mysterious power, many can testify. Notice the testimony of the Algiers woman converted from Mohammedanism to Christianity: She had been poisoned by her angry relatives to put her out of their way as being their religious disgrace. Having read that in *His Name* the dead should rise, and if a Christian should drink any deadly thing it should not hurt him, she began to call the Name. She persistently invoked it though the well-known symptoms of mortal hurt were increasing. Suddenly she felt as if a stream of pure water were flowing through her body. It kept on with mysterious swiftness, till every vestige of the poison was eliminated, and new life began pulsing through her being.

Who now is ignoring every disastrous state of affairs, and separating himself to the *One Name* by which the first Apostles wrought their miracles?

"And the throne had six steps." "And I saw in the right hand of him that sat upon the throne, a book."

"I wept much," wrote John the Apocalyptic seer, "because no man was found worthy to open and to read the book."

It is the book effecting by the mysterious writing on

its covering, the unlocking of the doors of limitation, and the uniting of whoever reads the outer writing to unstinted blessings.

"Seek ye out of the book of the Lord and read, no one of these shall fail." "For the deaf shall hear the words of the book, and the eyes of the blind shall see out of obscurity, and out of darkness."

Surely, the book in the right hand of the Lamb slain for the transgressions of the race, gives for its outer reading the Name or revealing so vitally insisted upon by the first Christian Apostles, and so ignored as to its mystical potency by the Christians of today. And the opening of its inner writing waits upon the faithful reading of its outer form. "Blessed is he that readeth"—"I will give him to eat of the hidden manna."

Faithful reading is always associated with understanding: "Whereby when ye read, ye may understand my knowledge in the mystery of Christ." "Whoso readeth, let him understand."

And understanding is mystical light. It is the light pent up and hidden in all men as in the first disciples of Jesus, and it is the glorious light of the Lord of the wide universe. "Then shall thy light break forth as the morning"—"And the Lord shall be unto thee an everlasting light."

The same law of light reigns for the unseen shining as for the manifest. Do not the men of science insist that the same light that illuminates the noonday sky is present in the darkened chamber? And that the bars of hiding being removed, the light within springs to meet the light without!

"Held our eyes no sunny sheen,
How could sunshine e'er be seen?

Dwelt there no divineness in us,
How could God's divineness win us?"

The breaking forth of the pent-up light, which is our hidden understanding, is associated with the bursting forth of our pent-up health: "Then shall thy light break forth as the morning, and thine health shall spring forth speedily."

It is associated with rising: "Unto you that fear my name, shall the Sun of righteousness arise, with healing in his wings,"—"And the gentiles shall come to thy light, and kings to the brightness of thy rising."

"Tarry ye in the city of Jerusalem," said the Risen Lord, "till ye be endued with power from on high"— "Ye shall receive power after that the Holy Ghost is come upon you"—"The Holy Ghost whom the Father will send in my name."

Then the obedient disciples tarried in Jerusalem. They tarried during the forty days of the Lord's ten appearances, plus the days to the fully come Pentecost, the Sixth day of Sivan, the feast of the harvest; coming together often, with one accord, to call upon the Spirit-imbuing name. On the sixth day of Sivan, the house where they were gathered trembled in the rushing Spirit they had invoked by their persistent calling. Tongues of mystic fire whispered to them the powerful mysteries. They were ready to greet the outside world with a new ministry.

Six is the number consecrated to final equipment for heavenly offices. No equipment transcends the effulgence that wakes in us the healing word, the tongue for speech with angels. Even the shadow of Peter, most name-distraught, caused joyous forgetfulness of pain.

Jerusalem is symbol of the Self. To tarry at the Self, invoking the Name that wakes magian majesty, inbreath immortalizing, medicine of God, is labor worthy the sons of men.

Judas of Judea set his eye toward secret cheatings, and inbreathed them with the Palestinian winds. He shut the gates between himself and the offered higher laws of success. Jesus of the same Judea set His vision toward the Author of success, and inbreathed the airs of Paradise. He opened the gates between Himself and unstinted transcendence. Judas hanged himself, and his name is full of suicide. Jesus rose to immortality, and His Name is full of life-giving, miracle-working energy; it opens the gates between man and his native kingship.

Choose this day the objective for your vision's ofttime gaze, and your calling's precious good. Be as hotly intent as Carlyle toward literary distinction, as Judas toward the bag of silver, his own bold business, as Jesus toward God. Whatever we look toward, we come into identification with.

He that seeks Me identifies with Me. He reigns with Me. He lives as My life, he strengthens as My strength, he understands as My understanding. What I Am he is. He calls upon My victorious Name, and whatsoever he does prospers, reminding mankind of My ever present, ever friendly, ever available Supremacy.

For I send the Healing Ghost, the Enwisdoming Breath, to him that calls My Miracle-Working Name Christ Jesus, bursting through which is the other Name, only known to them that invoke His Anointing Name.

Let us recognize the Jesus Christ Self of this planet as the Victorious One ever present.

Let us be insistent about it. Even distant nations must measure up to life-giving peace by our persistent attention toward the Radiant Solitary whom God hath set in their midst.

E. C. H.

VII

MINISTRY

The Angel of His Presence accompanies every man. It is his kingly Self. Two are ever in the field; one shall be taken, the other left.

I will send an angel before thee, said the voice of God to Moses. Because Moses knew this, he daringly followed his leader. Once the angel led Moses into Marah where waters were bitter and the people complained that Moses was a deceiver. But the angel pushed him to a tree to cast into the waters, and its absorbent qualities picked up the bitter globules leaving the waters limpid-sweet; then led him on to Elim where there were twelve wells of water and three score and ten palm trees.

This high leadership is every man's heritage. He need not fear dangerous days or vicious circumstances while he is aware that his angel goes before him, pleads his cause and defends him. High vision causes sense of nearness of the Highest. It is the closeness of the Ain Soph the Great Countenance of the Absolute, above thinking and above being, which the Hebrews called Angel of God, the Brahmins called Divine Self, or Stately Soul, the mythologists called Æsculapius, or Apollo, the Christians call Jesus Christ.

Some reality, or closeness of the Universal has always been the human insistence of mankind. To the earliest man *The Angel of God* was the happiest name for the closely near, ever ready Helper Whom man might throw his arms around and unto Whom he might cry, "I will not let thee go except thou bless me." Jacob got what he cried for; Moses got what he cried for; Samuel got what he cried for; Abraham Lincoln got what he cried for. Many today are getting what they cry for, putting their heads against the close presence of the All-Competent One Who hath surely said, "He shall cry unto me . . . (and) I will make him . . . higher than the kings of the earth."

"Who hath God so nigh unto them in all things . . . that we call upon him for?" asked Moses, who taught High Attention and the close Angel of the Presence to the Israelites, thousands of years ago.

Always the angel does wondrously, flies swiftly, mighty in strength, ministering life to the faint, help to the defeated, comfort to the despairing.

But we must notice the angel. Attention is the secret of the success of the combination. We combine with what we notice. We produce something worth while by combining with the Angel of God's Presence.

There are some unitings or combinings even among mankind that produce success, victory, splendid achievements. The unassuming Cadijah made the greatness of Mahomet to out-shine himself. The unassuming Stier made the genius of Pavlowa to transcend herself. The unassuming John made the speech of Peter to stir living sparkles through dead nerves. Many a great man owes even his ability to make money to the unpretentious mother or wife or child who lives in his house.

There are occult combinations that win outward honors. The science of the future will unveil the combinations. The Hebrews knew that Raphael was the angel of healing. Gabriel was the angel of comforting, Michael was the angel of good overpowering evil, Uriel was the angel of convincing doctrine.

That is, they named the miracles by their angelic names, as the Greeks named the healing that worked in their bodies *Æsculapius,* the strength that roused up in their nerves *Hercules,* the sleep that rested their bodily frames *Morpheus.*

Sleep was to the Greeks an entity. So also was health. Strength was a god by itself. All these were callable, that is, they could come by being called. The Christians united all the gods into one God, all the angelic ministries into one ministry, viz., the Sonship presence of the Universal Redeemer whose close visibility staid longer than any god of Greece or Rome was ever reported to have remained visible.

The Christian dispensation gathered the multitude of gods into one Lordship, Saviour visible for three years, then Saviour invisible but blessedly near for the rest of time. "Lo, I am with you alway," He said, which promise not one of the gods of mythology ever made, though forever and ever strength as a distant entity might be called near to stir the sinews, and health as a distant entity might be called near to charge the bodies of human beings.

These white gods had each his own ministry, but Christ Jesus had the combined ministry of all the gods. "All power is given unto me," He said. "Whatsoever ye shall ask in my name, I will do it."

All the gods with their several ministry were known to be representatives of some Universal Vast Vast. It was *chrysolite* stone of character to know how to be charged with the invigoration of some supernal deity and then, further, to cause the same invigoration to wake in one's neighbor. It was the ministering power of only a mysterious few who were called priests of Zeus, or priests of Æsculapius, or priests of Apollo, according to what god quality they could best awaken. We read of a priest of the Egyptians goddess Neith whose beautiful daughter was married to Joseph the Hebrew. Also in *The Acts of the Apostles* we read of a priest of Jupiter in Lystra who brought garlands to honor Paul and Barnabas for having healed the feet of a lame man crippled from birth, whom he as priest of Jupiter had not helped.

Those who had been sprinkled by Chronos the god of time did great things even in their old age. Evidently they would have said that Titian had unwittingly been sprinkled by Chronos for he painted *The Battle of Lepanto* at ninety-eight years of age; also that Michelangelo must have been sprinkled by Chronos for he was still painting great canvasses at eighty-nine. Whatever characteristic one could inhale from the ethers, or be sprinkled with from the gods consciously or unconsciously he was privileged to cause to rise up to some extent in his neighbors. He could pass it on as the contagion of his aura, or atmosphere, consciously or unconsciously, as was reported the cure of a love for intoxicating drinks passed on to an inebriate while he sat near a lover of the Free God in a public meeting.

Give attention to the Effulgent One like Parmenides and radiate effulgence. Gaze toward the glory of the Highest and ray forth some new aspiration. Seek Him that turneth the shadow of death into morning and ray forth reviving ethers that spiritualize even injured backbones. Be strong in the Lord and in the power of His might and ray forth peace and confidence in the midst of life's hardships. Be in love with some Truth of the

Self-Existent till the scent of its Rose Garden reaching thee, thy garments carry Balm of Gilead for pain. Is it not written, "I clothe My priests with salvation"? There is healing victory for every one in love with the Uplifting, Awakening Highest. His influence is irresistible. "Listen, O isles, unto me, and hearken ye people from far—I will also give thee for a light to the Gentiles, that thou mayest be my salvation unto the ends of the earth."

Beresford writes that some people's secret emanations or influences are very strong, so strong that they can be photographed. Let us note that only those who have been in love with Divine Rulership ray along divine influence, doing angelic ministries, mighty in strength, flying swiftly.

Calling unto the Name that smites the ethers into new activities creates new conditions. "Ye shall know that I am the Lord when I have wrought with you for my name's sake." "Therefore, turn yourselves and live ye—Turn others and live ye," calling upon My Name.

There is a Miracle-Working Presence. It is man's privilege and obligation to make identification with the Miracle-Working Presence, till he himself is a miracle-working presence, spilling over with new radiations as the opened flower spills over with new perfumes. All miracle workings are angelic ministerings.

It is well to notice *what* we are oftenest visioning toward, as it accounts for our personal conditions and our atmospheric influences. Tell the vivisectionist that his own bodily anguish hurries toward him, and that his subtle personal influence is pain breeding. Tell the worshipper of the High Redeemer that the liberty of the Sons of God hurries toward him undoing all sly contagions, and that his subtle atmosphere is pain-undoing.

Give the whole world the message, how that on his golden bed Solomon with eyes tightly closed gazing out over his unhappy subjects became so unhappy his groanings could be heard afar off; and how the greatest surgeon of Italy on his last bed of torture, with the instructions of another world touching his mind, declared that he must expect that to happen to himself which he had caused to happen to others.

The vision of man is persistent. It records. And it must formulate somewhere *nolens volens.* "Look unto Me and be ye saved." "Thine eyes shall see the king in his beauty—and the inhabitants shall not say, I am sick."

Isaiah writes of priests and prophets who by erring in vision stumble in judgment. He deplores their covenanting with their own unprospering imaginations. He insists that in this way they cause their people to dwell in the paths of destruction. But the golden thread of a healing doctrine runs all the way through his denunciatory messages. He is teaching the Hebrew prophets the curative ministry of the *high watch:* "Behold, your God, He will come and save you."

Notice how all inspired writers put the vision before the judgment; and the outward conditions of the people's experiencing, as results following judgment: "Where there is no (saving) vision the people perish."

Man's inward visional direction creates his judgments, or mentals; mentals then translate into manifest affairs and manifest bodies. Mentals unvitalized by high vision are but compoundings with phenomena that never get anywhere. "Canst thou by (such) thinking make one hair white or black?"

It is the glory of Mystical Science that its fundamental and first instruction enjoins the exaltation of the inward visional sense, in order that exalted thoughts may formulate and immortal bodies and joyous affairs be made manifest.

David was teaching according to mystical law when, looking up, he talked to the Author of the Wooing Edict, "Look unto Me:" "Thine, O Lord, is the greatness, and the power, and the glory and the victory, and the majesty . . . Thou art exalted as head above all . . . Both riches and honor come of thee . . . Keep this forever in the imagination of the thoughts of the heart of thy people."

The Greeks were clutching at higher things than their minds when they glanced up to Æsculapius the white god above them, who caught and held them tightly and performed even surgical operations as god of healing ever ready world without end, the sweet god's envoy.

The new education that is coming slowly to the front, tells by one or another mode of impartment, that whatever is oftenest viewed with the inner eye reveals its secrets and hands out its gifts. On this principle a certain choir master taught his boys to remember how the notes looked on the staff, and it was remarkable with what accuracy the boys sang the notes when after some days they came to voice practice. The notes had handed out their best to their attentive onlookers. Mozart lifted his inward gaze again and again to the sky's angelic choirs, and heavenly strains gave him mental musical ecstasy.

The Parsees taught that every man has a rich boon

destined for him, but that hardly ever any man receives his boon; he repels it, instead. Ostensibly the Parsee gives no hint of how a man may draw the boon destined for him; but going deeper into the secret doctrine threading its way through the Parsee, as through Hebrew writings, we find the sweet urging and the plain direction to set our vision high toward Him that turneth the shadow of death into morning: Every good and perfect boon cometh from above.

Some people by untaught fixedness of inward attention have occult vitality in visioning. They hasten their objective points to be plainly extant in the world of nature at large and in the world of their own human experience. W. T. Stead described the shipwreck of a White Star liner on an iceberg at sea several years before he experienced it. This was not a very prompt arrival, but J. A. Bartlett's vision of the Omnipresence of One Flawless Spirit caused a lame man sitting near by to suddenly throw down his crutches and walk with flawless freedom.

"I have said, Behold me, behold me . . . I have spread out my hands all the day unto a rebellious people," saith the Lord by Isaiah. "They have turned their back unto me and not their face," saith the Lord by Jeremiah.

Occult vitality, or swift formulating vigour is cultivated by steadfastness of attention to the High Helper, the Supernal Giver, who evermore is saying, "Look unto Me,"—"I restore"—"I help"—"I instruct"—Is there anything too hard for me?"

There are objective points to focus the attention toward that which Isaiah calls covenanting with death. Peter the hermit set his inner eye toward the Holy Sepulchre in Jerusalem and with impassioned descriptions drew the steadfast attention of millions of men, women and children toward it as their rightful possession. They endured as seeing an invisible sepulchre and nearly all fell into sepulchres of their own while journeying toward the eloquently described far distant Sepulchre of Jerusalem. How wonderful would have been the hermit's record had he used his hot eloquence to invite his people's attention to the Lover above, ever calling to them, "Seek ye me, and ye shall live." "Look unto me and live." "The way of life is above to the wise."

Why has no eloquence described the Countenance of the over-looking Father in Whose light we may breathe the light that infuses with the untellable life of the splendid Eternal?

The mother says, "You will catch cold." She does not say, "You will catch health." As the child has native identifying speed, he soon exhibits the objective of his mother's errant vision with the threatening name *cold.* If the little Ludovico could tell what page in a book his mother was looking at, even when he was not near her, why could not a child catch his mother's fears or her foolishness, why even could he not catch her fearlessness or her wisdom?

It was from persistent attention toward the All-Wise and Ever-Living Jesus Christ that Tertullian became the creator of Christian Latin literature, and was given to know the ministering assurance that "We shall all go on forever being the same persons we now are, and shall so continue forever clothed upon with the peculiar substance of immortality."

It was by observing the disasters in the earth that the unministering Epicureans contended that "God is inert, and a non-entity in human affairs."

Christianity came after Epicureanism, teaching that
the kingdom of God cometh not by observation of earth's
disasters; that we must "Look up to the fields white for
the harvest"—"To the God that giveth to all men liberally"—And God shall wipe away all tears."

True mysticism in every age and every land calls
the spreading appearances of earth *mirage and delusion.*
What inspirations, what instructions, can possibly arrive
as the results of much study of mirage and delusion?
What noble cures, what divine immuneness from neces-
sity for cure have arrived, or ever will arrive from
the observations of men in the torture chambers of vivi-
section? On the contrary, is it not written that "He
that killeth with the sword must be killed with the
sword?"

True mysticism teaches a practice that results in
cure for all its devotees, and exposes to them the immune offspring of Jehovah-All-Peace on every side.
He that looketh toward Me doth everywhere behold Me.
"I the Lord will hasten it."

In the eighteenth psalm David proclaims that he
called upon the Lord, and was delivered from his enemies and his wisdom candle was lighted. Solomon, his
son, explains for us that David's candle was the Spirit
of the Lord shining upon him: "The spirit of man is
the candle of the Lord."

Job in his identification with affliction lamented
the lost days when he had seen the candle of the Lord
shining on his head, which candle he himself had hidden
by much gazing toward suffering.

The secret of a man is his candle of lordship, his
spirit of wisdom. Dr. Arnold of Rugby studied every

class lesson faithfully in order that he might bring out the responsive intelligence—the candle of lordship common to the class—the Solitary that God had set in that little family. To his lasting honor, by addressing Responsive Intelligence, he brought forth leaders of men, statesmen, presidents.

An instructor must bring forth lordship of some sort or he has not accomplished anything. "He has gained nothing who has not gained the Soul," proclaims the Vendidad, mystically teaching men how to rise free from the clutches of defeat as stalwart lords over evil by ofttime lifting up attention toward the High Redeemer, Whose way upon the earth is the saving health of the nations.

Soul, Spirit, Light, Wisdom, are all names of the Responsive Intelligence filling the universe, ready to break forth everywhere by recognition. "With right glance and right speech man may superintend the universe, animate and inanimate."

"And the seventh day he (the Lord) called unto Moses." He responded plainly to the speeches of praise Moses had made unto Him. And He taught Moses how to build an ark-symbol, which should to the Hebrews be sacrosanct testimony forever that the Lord Jehovah does dwell among men, their Leader, Defender, Provider, Inspirer.

Praise of the ever over-looking, ever-fronting Provider, Defender, always meets with response. Was it not promised of old that the sceptre should never depart from Judah? Judah means praise. Praise of earth's bountiful productiveness has caused the earth to teem with plenty. Praise of gold has hurried gold in rich masses to show itself to praiseful gold seekers. Praise of labor has overrun the globe with laborers planning world conquest.

Recognition is a form of praise. Description is a form of praise. Even to describe what we do not like magnifies its importance and spreads forth its capabilities. "Mine eyes are ever toward the Lord," said David, and "my mouth shall not transgress."

Did not Fechner discover that Responsive Intelligence looked upward toward him through the earth, as he called it "Angel Mother"? Did not Simon Magus experience earth's levitating strength, when, by praising Responsive Intelligence as upward bearing strength he buoyantly rose high into the air? Did not Iamblichus draw the distant eagles from their crags to come near to him, by speaking firmly and confidently to the Responsive Intelligence shining through the eagles? Does not Bjerregaard declare that the whole earth is awaiting orders to change its perpetual destructiveness?

"Let your speech be seasoned with salt," said Paul, by which he meant *stern confidence*. We learn stern confidence by facing the Author of stern confidence, speaking praisefully to the High Original, Whose word changes not, Whose finished works, instinct with agreement that firm praise shall surely make manifest, endure forever and alter not. Concerning the work of My hands offer praise and thanksgiving. The sceptor shall never depart from praise.

"I send to you Epaphroditus," writes Paul; "He was sick unto death; receive him therefore in the Lord, and hold such in reputation" (or, "honor such") .

Paul may not have known how near he was bordering to Hinduism while saying they must receive such as Epaphroditus in the Lord and give them good descriptions; but the word of the mystic Hindu had for centuries before Paul's time been, that, "He that beholdeth all creatures seated in Me, shall behold Me on all sides."

"Behold Me, Behold Me," this is the never-ceasing Heavenly Edict. "Whoso offereth praise glorifieth Me;" this is the law of magnifying by description.

"Let none of you imagine evil," said Zechariah the prophet who had understanding in the seeing of God. Zechariah might have detailed the sickening imaginings of the mother, the death-dealing imaginings of Peter the hermit, the defeat-bringing imaginings of people like Cleopatra at Actium who, when in the midst of victory suddenly lowered her gaze and brought defeat. But Zechariah specified nothing in particular. He gathered all false imaginings against ourselves and our fellowmen under one head, and warned mankind forever to speak only truth to his neighbor: "Let every man speak truth to his neighbor, let none of you imagine evil in his heart against his neighbor."

What is truth? High praise of the lordship, the shining wisdom, the divine wholeness of every man, woman, child, animal, tree, plant, stone, star, throughout all the near and far stretching expanses, visible and invisible—this is truth. For back of each visible is divine reality. In the Hindu Lanka Vetara we read, "What seems external exists not at all."

What made the Christian Apostles so mighty in works? Their secret visional direction, forming right judgments quickly, making sudden outward demonstrations, being certain of a Countenance shining as the sun, raiment white as the light as the only ever facing Reality.

"I am determined not to know anything among you, save Jesus Christ," said Paul to the Corinthians. The Corinthians were famous for their strange diseases, but so long as the Lord Jesus Christ was standing up in the midst of them why should Paul set his eyes on strange diseases?

"After six days Jesus . . . was transfigured before them, and his face did shine as the sun, and his raiment was white as the light—And when they had lifted up their eyes they saw no man, save Jesus only."

The time always was and is now that the divine Self, the shining Spirit, the Jesus Christ Lordship stands ever before us, but only by having had our eyes lighted by lifting them to the Ain Soph the great Countenance of the Absolute, can we see face to face who our neighbor is in the sight of the over-seeing Father, and give our neighbor the honor due him as shining Spirit, flawless Lordship.

This was the secret of the wonder-working early Christians. He who had said, "He that hath seen me hath seen the Father," was the perpetual objective to their inner eye. He was to them the Angel of God's Presence, identical in office and authority with the Father. His promise to be with all men to the end of the dispensation of lower imaginations was a reality to them, as it is now to all who believe His words, "Lo, I am with you alway."

There always comes before us the Angel of God's Presence whenever any man, woman or child appears. "I am the truth," he is saying—"Judge not according to the appearance." It requires that strength of character called the *seventh stone* to answer, "I will not judge according to appearance. Noticing only the Angel of the Presence I can firmly declare, "I know you as free, wise, immortal Christ Jesus here present; neither sin, sickness, nor death, can touch you."

Once having started on this truth, the glory of truth so quickens our hearts that to us as to the upward visioned Apostles, there is no man save Jesus only, with face shining as the sun, and raiment white as the light. Nobody, even a Corinthian-plastered enemy, can resist the praise of his victorious free Spirit, his candle of Lordship, his Angel of God's Presence, his own real Self. He has to drop all his delusions, his diseases, his griefs, his misfortunes, even his death. Two are ever in the field. Choose ye.

"Go, I pray thee, Joseph," said Israel, "and see the peace of thy brethren." And Joseph went out into the world seeing only peace, the Jehovah Shalom ever present, the Lord of peace, and speaking only peace to his eleven brethren. From the standpoint of the unenlightened eye the eleven brethren were liars, thieves, murderers, disease-threatened hypocrites. But Joseph on about his father's business, saw peace, spoke peace, went forward toward the light of peace, till the eleven fell down before him transformed into lovers, friends, light bearers to the world.

The bringing forth of the Jesus Christ type, spiritually bold and executive, is the fruition toward which all religions and philosophies are bent.

The recognition, the acknowledgment of a Christ Jesus in the universe, focuses the inward attention to the greatest character-formulating objective the universe holds. It acts like heat on invisible ink. Mary had been with the inner eye looking steadfastly to her Lord as above in heaven, till He stood beside her, even while she with outward eyes beheld Him as the gardener. He called to her as He had called to Moses. Then with opened eyes she saw Him, and not the gardener. "The secret of the Lord is with them that fear him; and he will shew them his covenant, to make them know it."

Some later day philosophers tell of "The Soul, the Spirit, rising up in wrath against the natural order that denies its autonomy;" and of its "defying the principalities and powers that would brutalize it." But the Hebrew sacred teachings declare that Soul, in calm majesty, defying nothing, doth truly wait on God. "Nothing can injure the immortal principle of the Soul," wrote the watchful illuminati among the far past Theophoroi.

He Who is the Strong Son of God, standing ever in our neighbor's stead, manifests as health and strength in mankind, when glorious truth is spoken to Him as The Wonderful facing us everywhere. Nothing that we praise the Omnipotent Christ as being shall fail to manifest somewhere, somehow, among the sons of men. Why should we speak to Him, free Soul, free Spirit that He is, as rising up in wrath, or fighting with defiant zeal to hold His own among us?

Sometimes we read in writings of influential men, that God suffers at sight of our wickedness, and is distressed at our ignorance and stubbornness. For the moment the writers seem to be forgetting that "God is of purer eyes than to behold iniquity." By such teachings concerning suffering and distress on high, they turn the eyes of their docile people to an inadequate and despairing Deity who needs help and encouragement from us. They must study again the written inspirations of men in their diviner visions, when, as the Greek high watcher averred, they are "above their proper wits," unhypnotized by outward observations. There they will read of the High and Lofty One inhabiting Eternity, Who healeth the broken in heart, Who bindeth up their wounds, Who teacheth men what before they knew not, Who exalteth by His power, great in counsel and mighty to work, Who cannot look on evil.

"The Lord will raise up a prophet from the midst of thee, of thy brethren; unto him shall ye hearken," was the Mosaic prophecy. "And Philip findeth Nathaniel, and saith unto him, We have found him, of Whom Moses in the law and the prophets did write."

"And in his name shall the Gentiles trust." "And he shall show judgment to the Gentiles." "Watch ye therefore—to stand before the Son of man." It is necessary to right accomplishment that we have a definite goal to reach, a character standard to emulate. Failing a unital standard of finished high achievement, whole colleges of men have gone on year after year without graduating a single man of high achievement.

It is not from gaze toward the inchoate mass of men as brotherhood, or the mass of women as sisterhood, or the mass of labor as leaderhood, that we catch the sound of a distinct moral tone, or discover a pivotal health centre shedding forth resistless health. It is from gaze toward one victorious, indestructible Minister of the Almighty Original ever standing in our midst, saying, "Lo, I am with you alway," that we catch the keynote to victorious ministry.

He has gained nothing who has not gained sight of the ministering Solitary Whom God hath set in the family of earth.

Joshua saw the Lord Jehovah standing up in Jericho; and moon-worshipping Jericho became a school of the prophets of Jehovah. "As the Lord God of Israel liveth before whom I stand," said Elijah, not noticing Ahab and Jezebel and their warrior-strengthened animosity; and Elijah was independent of the angry king and his army of haters.

"In a very little while Lebanon shall be a fruitful field," wrote Isaiah with figurative profundity. *Lebanon* stands for the Soul type, the Jesus Christ standard, Strong Son of God, Everlasting Friend, always with us, asking only recognition by praiseful description to be promptly fruitful after His own pattern, till earth's cities and plains are filled with men fulfilling untellable triumphs by new laws discoverable only through association with an altogether triumphant Prototype.

"I turned to see the voice: and being turned," reports John the Revelator, "I saw seven golden candlesticks. And in the midst of the seven candlesticks One like unto the Son of man."

Candlesticks are definite divine messages. To know the seven definite messages of Pure Mysticism in their straight line of revealment, as decipherable through the many insulating or mistaken injunctions of the Sacred Books of all time, is to come like John, face to face subjectively with One All-God, Whose objective reality makes haste as the fulfillment of inward certainty. For, except what we recognize subjectively, has its counterpart outwardly, we have not consummated our convictions. Inward conviction is not alive till it demonstrates on the human and manifest blackboard of daily experience.

Make a note of John's final hardy exhibitions of what he saw on the Isle of Patmos, with inner vision unrelated to the harsh passing Patmos days: Men denied the truth on which his faith rested, and with railing and malignant temper disputed his authority; but Polycarp, Ignatius and Papias rose up ready for martyrdom for the truth of his preaching. Boiling oil had not power to hurt his Christ-imbued body; even robbers turned glad Christian healers at the sound of his confident word, "Thy Soul prospereth." At Ephesus, where he had been maligned and denied, he was carried into the Christian assemblies bearing on his brow a plate of gold with the sacred Name engraved on it; and he was named "Apostle of Love," not as a character feminine, yielding, softly benevolent, but as one who recognized the Lord Christ ever present as the only reality of every man everywhere.

Jacob saw the same Son of man, and called Him "Angel," and "God," and keeping his vision steadfastly set toward the Angel, he blessed with effectual blessing all his children and grandchildren to the present generation, and to all the mysterious generations hastening along. The Angel of His Presence saved them—in His love He redeemed them.

And the seventh foundation stone of character (the heavenly Jerusalem) is *chrysolite*. This was one of the symbolic visions of John toward the Son of man merging into manifestation. The chrysolite is the golden stone symbolic of right communication—the tongue of the wise that is health. Does not Solomon say that he that speaketh truth showeth forth righteousness, and that the lips of truth shall be established forever, and that the words of the pure are pleasant words, winning the Soul—the divine Selfhood, to show forth?

The divine Selfhood always accompanies every man. It stands with him, back of him, or near him, instead of him, uncontaminated by his human mistakes; indestructible though he himself appears to be broken; omnipotent though outwardly he appears weak; poised, wise, joyous, though uncertainty, stupidity and grief are now his manifestations. Who is there determined not to judge according to appearance, but to judge according to the Selfhood of every child of earth, praising the ever present upright man whose right is dominion?

Those willing like Paul at Corinth to know no man save the Jesus Christ Self only, touch the seventh stone—the golden speech of right association. "Two shall be in the field," said Jesus, "one shall be taken, the other left."

"Choose ye this day whom ye will serve"—by description. For all description, silent or audible, brings to outward exhibition; speedily, if the inward conviction is alive, slowly, if the inward conviction is not yet vivified by persistent attention. The downward watch calls for descriptions of death and defeat, disease and dementia, the shadow system, insubstantial, untrue, ever waiting the message of its own undoing.

True descriptions are the New Gospel. Praise the true Self, is the Right Message. Do not wail that little false notions, little errors are always creeping into your mind, though truly you want only God. Do not complain that your body is not sound, though you really love the Highest. Can you not see that your vision is on the terrifying errors creeping in, and on the body's non-refreshment, rather than on the "bad man's Deliverer," "the Lord that healeth?" The Sacred Edict calls for looking away from error and from human brokenness.

What shining conversationalists and clever describers of human conditions are among us! What gospels of healing they might be turning their splendidly equipped tongues into, by praising the free, wise, immortal Son of God standing up Solitary and Glorious in the family of earth! Every listener would be enchanted with his own divinity and forget forever his low estate of ignorance and pain. But even if the learned and eloquent of today refuse their privilege of chrysolite speech, the "Desire of all Nations" shall surely come, as response to some voice of praise of Reality, as the prophet Haggai foresaw.

Though the people were tired of Aristides for talking so much about Æsculapius and Minerva, his undaunted speech brought Æsculapius and Minerva into his own plain sight, with their ministry in his behalf, one to teach him statecraft for the people's benefit, the other to bring him health to continue helping the people.

Though they were tired of Paul for teaching Christ Jesus, yet by persistent praise of the "Desire of all Nations" Paul raised their dead for them, saved their proud officials from drowning, and cured their people of life-long maladies.

"Thou . . . hast kept my word, and hast not denied my name—and I will make them to know that I have loved thee."

"But in the days of the voice of the seventh angel, when he shall begin to sound, the mystery of God should be finished. And the angel said unto me, Thou must prophesy again before many peoples, and nations, and tongues, and kings." In his tenth chapter John tells the whole story of the Universal Christ ever present for manifestation. To prophesy is to proclaim truth, the same today as yesterday, and the same tomorrow as today. The eternal verities have no tense.

The first six angels, or announcements, insist on our looking above to the High and Lofty One inhabiting Eternity, King of Kings and Lord of Lords. The seventh angel declares that with the inbreath of the Victorious Name, the speaking tongue and the thinking speech are newly alive. The Victorious Name awakes vital breath and vital prophesies.

The old injunction of the magi to animate each particular life from the Universal Life is native activity to one who has caught the messages of John's mystically heard six angels. The seventh angel's voice telling us to preach the truth of the ever present Son of God coming before us whenever our neighbor faces us, is the call to us to speak to free Life till free Life shines forth.

Philip of Macedon kept a servitor near him whose business it was to be often saying in his ear, "Philip, remember thou art but mortal!" This worked so on the secret springs of Philip's outward activities that he rushed downward into assassination at forty-six years of age. Israel of old kept Ezekiel often reiterating what Israel should experience in the distant future. Ezekiel told Israel over and over that seeming national death should finally eventuate in upbuilding and regeneration. A captive himself on the banks of the Chebar, he proclaimed a far off future liberty for Israel by a new Law, or Torah. Futurity worked so on the secret springs of the Israelites that they are still waiting regeneration by a new Torah.

So under the Jewish mesmeric descriptions of futurity were the early Christians, that Paul said they would not accept deliverance from martyrdom, hoping for a better resurrection later on. And Paul, under futurity's spell, said that God Himself foresaw that the martyrs would not have been made perfect by suffering if they had not received the promises of futurity. Paul often forgot that "today is the day of salvation."

He who has awakened his speech by the Holy Ghost that comes by the Mystic Name, talks only to the awakened. He sees the Master Spirit wherever he looks, and he describes Victorious Spirit with the bold tongue of unhushable conviction. He looks to the Christ Jesus Spirit of the man who hates, and comrades with the Lover only; then nobody can find that man of hate. He talks to the Christ Jesus Spirit of the man who offers to "die for the Right," and he converses only with the Indestructible Master for the Right; then nobody forevermore can think of "dying for the Right." It is unthinkable. He thinks only that Right is Its own Life victoriously unkillable.

Right vision so regenerates judgment that the tongue and bodily activities regenerate in all directions. This is not more than Joseph did with seeing Peace, nor more than John did with seeing the Angel of the Presence, who told him not to do obeisance, but to know him as Mighty Brother, Master for Right, without hurting or being hurt.

As far as John went in his convictions born of his vision he demonstrated among men. As much as he visioned distinctly he proved. To him the seventh angel was the proclamation of the complete—the finished: "And the seventh angel poured out his vial into the air. And there came a great voice out of heaven, saying, It is done."

And the servants informed the nobleman: Yesterday at the seventh hour the fever left him. So then the father knew that it was the same hour in which Jesus had said unto him, "Go thy way, thy son liveth." The words of Jesus had been spoken to unfevered free Spirit, and outward conditions had tallied with unfevered free Spirit.

John prayed that his neighbors might prosper and be in health as he saw their soul prospering, and he declared that the greatest joy he had was hearing that his converts walked in truth, doing faithfully to their brethren and to strangers.

"In the seventh place the Lord imparted them speech, an interpreter of the cognitions," writes the Son of Sirach, in Ecclesiasticus 17:5. Seven is thus implied when we read that "The Lord hath brought forth our righteousness; come and let us declare in Zion the work of the Lord." "And the Lord shall bless thee out of Zion, and thou shalt see the good of Jerusalem all the days of thy life." "Zion" is man's divinity-Self, the upright that hath dominion. And the greatest blessing we can bespeak Jerusalem (our neighbor) is his divine freedom, as all sacred lore and law insist:

"In that day shall thy mouth be opened to the free." "In the seventh year he shall go out free." "Ye had done right . . . in proclaiming liberty every man to his neighbor." "That thou mayest say to the prisoners, Go forth; to them that are in darkness, shew yourselves."

Always high vision stirs right speech, silent or

audible; and right speech, silent or audible, stirs right convictions, and the accomplishment of the seemingly impossible is the outcome.

The upward watch of Benezet the shepherd of Savoy stirred in him such vital conviction of the presence of a living Christ that the Christ spoke plainly to him, giving him orders to build a bridge across the river Rhone. The public monuments of Avignon attest that Benezet did build the bridge which even Charlemagne would not undertake.

The upward watch of Dorothy of Brixton stirred in her such vital conviction that the same Lord Christ spoke plainly to her, telling her to rise off the bed of death and show herself strong, sound and sane. And she rose up strong, sound and sane. These both heard the same Responsive Power speaking to them that Moses heard "in the seventh day," as we read, Exodus 24:16.

Everywhere majestic Divinity faces us. Speaking to majestic Divinity stirs celestial instinction. Celestial instinction is that secret spring to right conditions which all the world is seeking. Instinction, vital conviction, is sometimes called the will. Has it not been declared that nothing ever really wins the will from its native bent? But *high watch* wins the will to work the kingly ministry of the flawless Deliverer standing up in the earth. "Did not we cast three men bound into the midst of the fire? Lo, I see four men, and the form of the fourth is like to the Son of God." And seeing thus the *Saving Other,* the king found no hurt in the three men of the fiery furnace in the plain of Dura. Was not that an exhibition of the worthwhileness of sighting the Son of God, the Radiant One ever present?

"And Enoch also, the seventh from Adam prophe-

sied of these, saying, Behold, the Lord cometh." Enoch
was mentioned of old as a type of perfected humanity.
He was described as one living a prophetic life and
having constant converse with the unseen world. He
taught our true human existence in glory, and the resurrection of our body in beauty. The voice of early ec-
clesiastical tradition regarded Enoch as one of the two
witnesses John mentions in Revelations, eleventh chapter, to whom should be given power to prophesy twelve
hundred and sixty days in the midst of world-wide unbelief. Enoch was the Scriptural forerunner of all who
look out over the earth toward the Angel of God's Presence, the Christ Jesus of Mankind, harmonious, wise,
immortal, and keep on describing the Saving One
though war, ignorance and destruction throw their shadows across the shining majesty.

Out of John's vision we gather the straight instruction that the word concerning the Lord's presence among
us shall prove itself to be the most mighty word, for
He shall surely be seen taking to Himself His great
power and reigning visibly among all the kingdoms of
the world. Why not, since what we persist in praising
must surely stand forth?

Ezekiel once had this mystic vision of the Son of
Man, the Lord-Self among the seeming dry bones in
the valley, as he prophesied divided Judah and Joseph
as soon to be one great kingdom alive with the word.
His speech to the dry bones has the vitality of some
unseen ocean's morning winds. He is a phonograph
of the Voice of the Master Builder: "O ye dry bones,
hear the word of the Lord. Thus saith the Lord God
unto these bones, Behold, I will cause breath to enter
into you, and ye shall live. And I will lay sinews upon

you, and will bring up flesh upon you and cover you
with skin; and put breath in you, and ye shall know that
I am the Lord."

Benvenuto Cellini sent his inner eye out lowering
down into distant space and caught the form of Charon.
He painted the form and features of Charon on plaques
and vases and shields, till he feared Charon as a dangerous entity. Martin Luther sent his lowering inner
gaze into dread space and described Satan, till Satan
became real to him—so real that at this very day an ink
stain on a castle wall shows where Luther hurled an
inkwell at the prince of darkness his mother had so often
described.

And to this day other human beings are still setting
their eyes toward menacings that do not menace, when,
like Enoch and Ezekiel in their great moments, they
might be viewing the ever present King in His beauty,
to the transformation of human existence into glory,
and the revival of the human form into beauty and
vigour. For at the sounding of the seventh angel's message, great voices shout that "The kingdoms of this
world are become the kingdoms of our Lord and His
Christ, and He shall reign for ever and ever," transforming war into loving kindness, and anger into chants
of praises of ransomed man.

To gloom or glory is the tend of the inner eye.
And all human beings and all events of human existence
are gloomed or glorified to us by the direction of our
inner eye. Luther at one period saw Satan overshadowing everywhere. Every direction then became so
gloomed to Luther that Catharine asked him if he
thought God might be dead.

Choose not to set the inner eye toward menace.

Choose to set it toward the prize of the Lord Christ's healing face ever looking toward us—Strength of the nations, Joy of the world.

> "Here eyes do regard you in eternity's stillness;
> Choose well, your choice is brief and yet endless."

> "Let a man contend to the uttermost
> For his life's set prize, let be what it will."

In one of Isaiah's dreams, his tongue was touched with prophetic fire from off the altar, because his eyes had seen the King in His beauty. And all sorrow fled from before Isaiah's face, and sighings he could not hear, for the songs of the unshadowed were in his ears, and the sight of the lame man leaping, and the weak hands strengthening, filled him altogether with joy.

If Isaiah with his will to concentrate all the energies of his being to the salvation of mankind from war and pain, had had the science of making visions immediately tangible, instead of always throwing them into futurity, Israel and Judah would have made manifest the sons of peace and healing in all directions, as Isaiah's delectable demonstrations of the *radio in extenso* of a finished vision firmly and sternly declared. "Believe his prophets, so shall you prosper," counseled Jehoshaphat the reformer; and the Hebrews believed in the prophets of futurity and prospered. Let now the world believe in the prophets of what already *is*, in the already presence of One Who redeemeth all life from threatenings for the seer's sake, that mankind may visibly prosper everywhere, and be in health, as Soul, the Only Reality of man, his Sonship to the Highest, already prospers.

"By the bright Soul's law learn to live;
    And if men thwart thee take no heed;
And if men slight thee have no care.
    Sing thou thy song and do thy deed,
Hope thou thy hope and pray thy prayer,
    And claim no crown this does not give."

Every spiritual law calls for steadfastness. There is but One Edict but we must steadfastly maintain obedience thereunto. Thus only are the practices signified by the voices of the angels, our own spontaneous practices. But having scientifically learned the voices of the angels, it is scientific for us to maintain them day by day under all the circumstances of daily living whether or no they have as yet proved themselves. Strike the iron till it is hot. Is it not reasonable to speak often praisefully and with firmness to the upright, ever-present Lord of Cure, if by so doing the radiance of the Sun of righteousness with healing in His wings shines forth, affecting all in our vicinity, as the wonderful Fourth affected Azariah, Mishael and Hananiah for the King of Babylon's sake?

"Man has the enunciative reason; if he makes use of this for what he ought, he will be guided into the choir of God" prophesied thrice-wise Hermes. "Do thou thy work betimes, and He shall reward thee in His time."

"My praise shall be of thee in the great congregation," said David; "That thy way may be known upon earth, thy saving health among all nations." Could any enunciation be more calculated than positive praise of the Lord ever facing us, to guide us into the choir of God? Why should we not joyfully practice the seventh angel's orders, "Prophesy again, before many peoples and nations, and tongues and kings." For "He healeth the broken in heart, and bindeth up their wounds. . . . His understanding is infinite."

"I will praise thee among much people." "On the harp will I praise thee." "Forasmuch as there is none like unto thee, O Lord; thou art great, and thy name is great in might." Let us take our eye off cheapness and incompetence, the wounded in life's battles, and the defeats of strife, and look often to the *Saving Other* awaiting our praises.

"He shall deliver the island of the innocent; and it is delivered by the pureness of thine hands." *Hands* are powers. No powers so pure as unqualified praise. But, notice, *He* shall deliver the island of the innocent. Some law as mysterious as that by which the trees leave forth, makes the unseen divinity of man, his island of innocence, kindle forth as health and newness of life, when the divinity is called Jesus Christ, the everlasting, ever-present, omnipotent Self, the Christ in you the hope of glory.

Note that David praises Him on the harp for the might of His name. We may never speak the name Jesus Christ. We may always speak the name Spirit, or God or Love, and describe Spirit, God, as Omnipresence, Omnipotence, Omniscience, and do quite great things by such praise, but there will be some nameless flavor lacking in the health and strength that come forth thereby. As the actinic ray in the sun is the secret of the sweetness of the grape, so Jesus Christ Who stands up in the universe as the one man having demonstrated the fullness of the Godhead bodily, is the human touch to the divinity of man. And being recognized as the health of man, there is a vital kindling forth as health and

happy vigour, fearless of death and misfortune, not translation but transfiguration for humanity.

"God hath given him a name which is above every
name." The preaching of the mysterious might of the
Name and the healing Presence of the Jesus Christ is the
risen doctrine. It is not the province of the risen doctrine
to insist on a verbatim formula of praise of the far radiating Presence. The recognition, the acknowledgment of a
pivotal character transcending all limitations, irresistibly
acting as invigorating energy, causes right words silent
or audible to formulate, sweeping their original meanings
into the affairs of the world.

"Now therefore go," said the voice of the Lord to
Moses, "and I will be with thy mouth and teach thee
what thou shalt say." Moses was slow of speech, but he
spoke and thought and wrote words that have caught the
ears and hearts and actions of millions of strong men
throughout many centuries. "The Spirit of the Lord
spake by me," said David, "and his word was in my
tongue." And David's words have been radiant with
beauty and vitality since ever he thought and wrote them
forth, kindled by the Spirit of God. For I "shall put my
spirit in you . . . and ye shall know," declared the Lord
through Ezekiel.

"I will give you a mouth and wisdom which all your
adversaries shall not be able to gainsay nor resist," said
the prophetic Christ to His first disciples. And they
were not able to resist the wisdom and the spirit by which
Stephen the miracle-worker spake.

The ever-present Jesus Christ is a resistless healing
and strengthening centre. How can we help speaking
the Name and describing Him as the burden bearer of
the world—the one Almighty Substance upbearing our

sick neighbor and his sickness—upbearing our unfortunate neighbor and his misfortune—upbearing our insane
neighbor and his insanities—upbearing our wicked neighbor and his wickedness—upbearing our palsied neighbor
and his palsy—upbearing our burdened neighbor and his
burdens—upbearing us and our obligation to help our
neighbor, till all the nations fear and tremble for all the
goodness and for all the prosperity the acknowledged
Son of Light procures them, as Jeremiah foretold in his
Chapter 33.

No address to our neighbor's mental constitution
(conscious or unconscious, so called) by the brave, bold
insistence of mental suggestions of freedom, peace,
strength, can equal the curative activity of the Son of
God acknowledged as upbearer, order bringer and healing peace. "It is he that giveth salvation to kings." "He
maketh doctrine to shine as the morning, and sendeth
forth her light afar off—and leaveth it to all ages forevermore."

It is the nature of the rose to shed forth perfume. It
has faced the sunshine that causes heavenly breaths to
smile forth from unscented leaf capsules:

> "How sweet the breath beneath the hills
> Of Sharon's dewy rose."

It is the nature of wine to refresh and revive. The
grape has faced the sunshine that wakes reviving wines.
It is the nature of man who faces the Countenance that
shineth as the sun in his strength to shed forth beautiful
descriptions that fit only the ever present Sun of righteousness with healing in His wings. And he who is full
to the brim of his being with silent descriptions of the
Lord of Glory now looking toward us sheds healing

perfume, curative breaths, reviving elixirs through even his garments. Did not Paul's aprons and handkerchiefs cure the sick? Did not Peter's shadow revive the faint? The Lord of life and health was their comrade. They kept high company. Not a new set of men and women came to them to show forth their unseen comradeship, but the same men and women came before them, clothed and in their right minds. "For thus saith the Lord, Behold I will extend peace . . . like a river, and the glory of the Gentiles like a flowing stream." "And ye shall know that I am the Lord, when I have wrought with you for my name's sake."

"And when he had opened the seventh seal, there was silence. . . . And I saw the seven angels which stood before God, and to them were given seven trumpets."

The seal of man's being first stirs to opening when he silently addresses the God that standeth in the congregation of the mighty, the shining One ever in our midst; and the seven trumpets are his when he boldly speaks aloud that the Jesus Christ of man is the only reality of man. The trumpets of God are the tongues of his prophets speaking in heaven-taught moments, when no complainings or denouncements caught from sights of evil spoil the music of their tones.

In the days of the voice of the seventh angel John was told that the book he must eat, which was the Name that should teach him all things, would be sweet in his mouth but bitter in the vital centres of his being. If John had not known by the third angel's speaking, that the wormwood of mind cure was the blessed bitterness of the book, and that the seventh angel's voice was only emphasizing the former prophecy, that his influence should be as a river of mind cure if he assimilated the

Mystic Name, he could not have understood the glorious gospel reiterated by the seventh angel's instructions.

But John understood that he was being told anew that his inward speech should relate him to the great Jesus Whose fame went throughout all Syria as healing the people, and throughout all Judea where the multitudes had only to touch the hem of His garment to stir with newness of life. John felt himself blessing countless generations by reason of comrading with the great central figure of cure and blessing. He felt the river of mind cure flowing forth from his own inward speech.

And the Son of Sirach prophesied, "Whoso feareth the Lord it shall go well with him." To *fear* is to keep the eye single. "If thine eye be single thy whole body shall be full of light" (wisdom). "The fear of the Lord is the beginning of wisdom."

The *Lord* is the resistless Law of the High Deliverer making manifest among mankind as the result of upward watch. Looking to the High and Lofty One inhabiting Eternity we bring to pass a new order of outward appearance to represent our high vision. It is always a masterful appearance. Nothing can hold opposition to it. If by our steadfast watch in obedience to the high mandate we suddenly see peace before our inward gaze, that peace will prove its lordship by quelling pain and discord. This is the lordship of peace. Jesus Christ ever standing before us is forever the Prince of Peace. His Name is the Mystic Name of the resistless lordship of peace. The Lord maketh it go well on all sides for them that recognize *Jehovah Shalom.*

> "Who by this vision splendid
> Is on his way attended,
> Great peace shall mark his way."

And the Son of Sirach further said that his right tongue, his new tongue, was his reward for praising the Lord, the Saviour with the Great Name. And this reward he called the time of the Seventh.

"And the Lord turned the captivity of Job when he prayed for his friends." And his friends were praying for him. Throughout all the mystic teachings of the divine law of cure, there is mention of prayer and praise in behalf of the brethren of the faith. Notice John's instruction, "Do faithfully whatsoever thou doest to the brethren, and to strangers."

Jesus began by sending His disciples out among strangers by two and two. "Two did build the house of Israel." "Two did put ten thousand to flight." "How should two put ten thousand to flight except the Lord had shut them up?" asks Moses, calling his people's attention to the miracle-working Lord. "If two of you shall agree on earth as touching anything that they shall ask, it shall be done for them," said Jesus; and, "It is also written in your law that the testimony of two witnesses is true."

Happy in our own day is everyone who has a comrade to go forth with him, keeping the high watch and praising the divine Self facing them, the Angel of God, the unweighted flawless Reality of every neighbor everywhere without respect of persons. They are surely like Peter and John at the Temple gate, speaking to the Jesus Christ Self of the impotent everywhere, to walk and praise their Lord. Zechariah, the seer of God, prophetically beheld two women with wings, bearing an ephah between them. And the wind of Divine Spirit was in their wings, and they bare the ephah to build for it an house in the land of Shinar, *hostility.*

Somewhere among the books earlier than Zechariah's time, it reads that two women together bearing the same message or ephah, could bring all the world to their doctrine. But surely the doctrine must be divine truth to be strong to hold out in the land of hostility (Shinar) to the day of the world's acceptance. Only divine truth can survive the Shinar hostility of downward visioned mankind. Two bearing the heavenly ephah of high vision and the praise that make up its ministry shall have speech with Responsive Intelligence, the Jesus Christ Self, standing free, wise, immortal, before them always. "Behold, He goeth before you," said the angel to the two Marys.

Job was one, and his friends together made another. Peter in the prison praying for his friends made one; his friends in an upper chamber praying for him made another. The great quarternity of sorrows, the dark d's of the downward looking of man's inward eye, disease, death, demonism, darkness, like a four-thicknessed veil, dissolved always before the heaven-sent *two*, the *two* standing before the Lord of the whole earth. "Even the devils are subject to us," reported astonished Peter.

The law of the way of the mighty miracle has never been abrogated. To all who read the words of the seventh angel, or the seventh chapter of the divine law of cure, all the miracles are free grace to the praiseful. Each reader makes the first of the two bearing right praises, the new ephah, the rest of those who love and agree with high praise of the Lord in the midst of us mighty to save make the other of the two bearing the ephah or the praise message.

It wakens a new and more spontaneous dealing with the flawless divinity of the distressed and ill, if we first look to the Divine One standing Solitary and Glorious in the midst of the people of the New Age, saying boldly to that Shining Presence: I see you transcending all human conditions, unweighted by matter, unshadowed by fear; free, flawless, triumphant. I look now to the Son of God, the Angel of God's Presence, the cure and peace and strength of all mankind. I look to Everlasting Life, shedding forth the radiance of the Everlasting Kingdom. All are gathered into life ever renewing, into health ever restoring, into wisdom ever brightening and rejoicing—gathered and upborne in beauty and love and might by Thee, O Free Spirit, victorious Christ Jesus facing me, with face shining as the Sun and raiment white as the light, transfiguring the whole earth with living triumph!

To address Responsive Reality in this way is to come into ready praise of the victorious divinity, the Angel of his Presence of each claimant for our help. Praise of the ever present Son of God distills as wine for the faint in the wilderness, sweet friendships to the lonely, blessed restorations to hearts sick of earthly hardships, for there is ever a baptism from high praise. It falls like gentle healing rains on human beings. It unites them to that which causes them to transcend themselves. Enduring as seeing the Invisible Solitary is greeting That which sees us, and the outcome is Health that rays forth Health.

> "The healing of Thy seamless dress
> Moves past the ways of men,
>   They feel Thee in life's throng and press
> And they are whole again."

By practicing the high directions of this Eighth Study as concerning everybody everywhere, and concerning every thought and every thing, we bless them all by undeceiving them all.

E. C. H.

VIII.

Ministry

The point of each chapter or lesson in Mystical Science is its practice point. The practice point of the fifth lesson for instance is not the acknowledgment of the Vicarious Jesus, but the Self recognition that He practiced and urged all mankind to practice, that each one might do something great in his own line as He Jesus of Nazareth had done in His line. His leadership in Self recognition was His credential to special acknowledgment and special praise on the part of all mankind. Self recognition follows recognition of the Angel of the Presence as Miracle-Working Nearness. "I will not let thee go except thou bless me" should be the Jacob cry of all the world, face to face as all the world is with its Jacob Deliverer.

The practice point of the sixth lesson is our own inspiration or inbreath of the Breath of Brahma. "I put my breath in you and ye shall live. I put my spirit in you and ye shall know." "Why O man will ye die, having power to partake of the breath of immortality?"

The practice point of the seventh lesson is recognition of the Angel of God's Presence as the original Self of every man, woman, child. Two are ever in the field, one should be taken the other left, as Jesus himself discovered.

To notice the unweighted, flawless, everlasting Self, or Angel of man is to notice the Christ Self, strong son of

God, lover divine. It produces a change in the bodily
and mental presence of anybody to notice his unweighted,
everlasting divinity with praiseful acknowledgments
silently spoken or audibly declared. "A right word how
good it is, who can measure the force of a right word."
"With right glance and with right speech man may superintend the universe, animate and inanimate." Notice
always how potent is a glance "Glance up often, so shall
thy life renew." "Look unto Me and live."

An eminent bacteriologist found that the indicator
in a delicate instrument moved back and forth with the
direction of the eye looking at it through a sighting slit.
The mystery of vision will some day be declared by science in such a way that mankind will know that to set
the eye toward the divinity Self of the neighbor is to find
its tangible beauty coming forth; and to sight toward
Deity is to experience the workings of Deity. "Seek ye
the Lord and his strength, Seek his face evermore," will
not be a beautiful sentence on a page, but a living fact
according to high science. "Deity onlooketh thee. Onlook
thou Deity. This, to thy salvation." Or, "Look unto Me
and be ye saved, all the ends of the earth."

To endure *as* seeing the invisible is to fetch it into
visibility. That means that we see *toward* what we cannot see till it arrives into our living experience.

A certain set of people choosing to see toward health,
wholeness, unspoiledness, declare that sick people are
made whole by their practice. Another set of people are
seeing toward riches, and riches come into their living
experience: "Seek ye first the kingdom of God, and his
righteousness and all these things shall be added unto
you," said Jesus. The whole and beautiful kingdom of
which he was speaking was noticed by Brahmins: "Above

this visible nature there exists another, unseen and eternal, which, when all things created perish, does not perish," they said. It was "The Perfect Land" looked toward by Egyptian seers of most ancient times. "The Archetypal world, the *Yesod*, the nourishment of all the worlds" wrote the seer of the Cabala. "The mother of Moses saw that he was a goodly child," reported Ezra, the illuminated priest of Babylon. And forth came the all-competent Moses, learned in all the wisdom of the Egyptians, a matchless warrior, a godlike law giver, a miracle worker unsurpassed. That mother must have had inner vision toward the flawless, victorious Angel of the Presence to have brought forth such a victorious man. It is written of him that he was a finished mathematician, the inventor of boats and engines, instruments of war and hydraulics; also the author of the accepted Egyptian hieroglyphics, ever leading by his lonely self the ascetic life that he might pursue high philosophical speculations and prophetic insights. Secular history presents him a noble example of the effect of noble vision,— so noble that critical intellectuals have not been able to believe that such a character ever really lived on this earth.

Whoever glances upward toward the Countenance that shines on his face begins to be lifted upward. Why should he not be lifted upward out of ignorance and feebleness altogether by ofttime returning the ever-seeing toward him of the Perfect Deity? Does not the hidden oak tree of the rotting pulp inside the splitting shell come up into another country and another breathing space by ofttime glancing upward toward a sun it sees not and untold homecoming for which its heart is ever dumbly yearning?

Mankind yearns for a country and a comradeship he cannot find except by enduring as seeing toward that city which hath no need of the sun or the moon to lighten it, for his own Beloved is the light thereof.

St. John of the Cross looked up toward a finished kingdom, and the furrows in the Monastery field were found to be miraculously plowed.

Even the most intellectual critics of mystical claims acknowledge that "the concepts of future creation are present in their completeness in the Eternal Now before being brought to birth in the material sphere." But they neglect to mention in such splendid assertions how to fetch to birth in the material or tangibly visible sphere the heavenly eternals already finished. For this we must look to Mystical Science. "And he brought him (Abram) forth abroad and said, Look now toward heaven, and I will give thee a son of her (Sarai) and she shall be a mother of nations; kings of people shall be of her."

The finished kingdom, the "Archetypal World" is forever wooing the sons of men to look toward it that they may find themselves and their environments blessed with supernal newness.

The stone was rolled away for the two Marys because their inner gaze was wooed heavenward. Lavater's gaze was upward toward the All-Knower, and his mathematical problems became easy. Martha's gaze was toward pans and brooms and bread for supper, but Mary's gaze was toward the Perfect Country, and Jesus said she had chosen the better part, or the visional direction that brought better results.

The Wooing Countenance shining into our faces from above was Agni to the Parsees of old. They declared that man must have been formed out of the eye of Agni, because his initial and compelling faculty was his eye faculty. Wherever his vision was directed, there his other faculties caught their sensations. This explains why the great Catholic Saints caught such tangible experiences. Lukardis of Ober-Weimer secretly gazing toward "some delicate invisible refection which the Convent could not afford, there came to her one day the most loving Infant bearing in his hand food, and begging her to eat it for his sake. She did so and was wonderfully strengthened." So writes Baron von Hugel, among other mentions of visible things appearing from invisibles with tangible outcomes, because of *sighting toward.*

Francis d'Assisi saw the invisible five wounds of an invisible Jesus on an invisible cross so perseveringly that he the visible Francis had the five wounds visibly present in his own visible hands, feet, and sides.

The point of the eighth lesson or study is the executiveness of persistence, perseverance. "Seek ye my face ofttime evermore."

"Persistence of vision" is a scientific term. It was called one-pointedness by the Brahmins. Man may behold the face of the Father in the Eight, wrote a lover of numbers, meaning that persistence in ofttime glancing lightly and quickly toward the Onlooking Deity, the God Countenance beneficently looking forever upon us, gives us the sense of friendship divine, effecting our awakening into our own flowering greatness. By persistent attention toward the Divinity Self, the Ageless Son of God, Plato found that "nothing can injure the immortal principle of the Soul."

By watching toward Apollo, the invisible god of good gifts, for nineteen years, the Greeks and Romans saw him visibly present and heard him promising happy

harvests and safe child-births. By much asking outward
toward some invisible informant a great chemist saw as
in a trance the form of a stranger who answered his question. By ofttime setting his inner eye toward the One
Invisible filling the Universe, Parmenides saw himself
as unified with that One so that the forces of nature de-
sisted from their wonted operations for his sake.

"After *eight* days again his disciples were within . . .
then came Jesus, the doors being shut, and stood in the
midst, and said, Peace be unto you." Everything that has
to do with the visible ministry of the invisible Jesus
blazing forever with the Christ Substance acts quickly,
for that is Reality already complete. The suffering Jesus
is an imagination of the heart and takes more time to
tangiblize, as witness all the bearers of the stigmata, so
long in manifesting the five wounds: It is high time for
us to take our inner gaze off the suffering Jesus and set
it toward the Risen Victorious, Christ-empowered Jesus.
"I if I be lifted up," He said. It is time the world took
its inner eye away from an angry God and set it toward
"Him who healeth all our diseases, who redeemeth the
life from destruction, who crowneth with loving kindness
and tender mercy."

It is time we took our eye off the grave and set it
toward Him who promiseth forever, "I will ransom them
from the power of the grave," "Seek ye my face and
live." Let us remember that the grave cannot demonstrate
our acquaintance with God; only "the living, the living
shall praise thee."

Even the ancients of Parsee India discovered that
watching things that spoil and die affected them with
spoilings, but ofttime upward glancing blotted out and

washed away the spoilings. "Lift up thine eyes." "I even
I am he that blotteth out thy transgressions for mine own
sake."

The point of the ministry of the eighth lesson, is "He
sent me to preach deliverance to the captives." All the
world is groaning for its rightful deliverance. Some are
groaning in the bondage of bodily pain. Some are groaning in the bondage of financial limitation. Some are
groaning in the bondage of mental inadequacy. Some
are groaning in the bondage of bodily inadequacy. It is
no use trying to get free by any other attempting than
upward face to face with Self-Existent, Untrammeled
"Ain Soph, Great Countenance of the Absolute, above
thinking and above being." "Thou has redeemed us out
of every kindred, and tongue, and people, and nation,"
acknowledged the high visioning John.

In the Dhammapada we are told that wisdom cometh
from above and the wise man casts off all shackles; that
right religion leads to escape from pain, and deliverance
from destruction. Also that the best doctrine is that which
removes pleasure and grief from the mind. As mind is
mirror of the vision it stands to old-fashioned reasoning
that visioning toward Free, Unattached, Unhindered
God the mind must be free. And as body follows mind's
agreements body must be free also.

"Lift up your eyes on high and behold who hath
created," said the heavenly voice to Isaiah. "Behold the
Lord God will come with strong hand, and his arm
shall rule for him." The coming of the "Lord" is near,
our own near angel of deliverance. Something mighty
to save is near at hand to do the saving. "The angel of
his presence saved them." Always the Universal Absolute

seems near as a strength beyond strength doing for us. Who notices the touch of the Almighty to save? "The hand of God hath touched me," said Job.

The Brahmins called the Wonderful-Competent our Real Self of ourself. The Hebrews called it the Angel of God's Presence. The Chaldeans called it the Stately Soul. The Christians called it the Christ Jesus ever present.

Some people are trying to show that Jesus of Nazareth did not claim to be the Promised Messiah, the Christ, the Saviour. But He did. Notice Him saying, "The Son of man came that he might give his life a ransom"—"I come not to judge . . . but to save the world." "The woman saith unto him, I know that Messias cometh, which is called Christ . . . Jesus saith unto her, I that speak unto thee am he."

They also declare that He did not teach the Trinity. But He did. Notice Him saying: "I and my Father are One." "The Holy Ghost whom the Father will send in my name, shall teach you all things." "Go ye, therefore, and teach all nations, baptizing them in the Name of the Father, and of the Son, and of the Holy Ghost."

They also proclaim that He did not urge any sacrifices or sacraments. But He did. Notice how He said, "If any man will come after me let him deny himself, and take up his cross" (the practice of denial of all that is not Free God). The ordinance of the Lord's supper was His sacramental ordinance to all time. "This do in remembrance of me," He said.

Keep the eye single to One Free Self-Existent God and you cannot help believing in Free Grace. You cannot help having a mind devoid of the pleasure and grief which worried the Brahmins and Buddhists. A new joy such as New Visioning gives causes the queer saying, "I believe hardships fall away of their own weight," in place of, "I am in bondage to hardships, I cannot cure myself."

Nobody is asked to cure himself of his particular bondage. He is asked to look up to Him who redeemeth the life from destruction, taking off every yoke. To Him whose only order has been, "Look unto Me, and be ye saved, all the ends of the earth." "The windows from on high are open, the earth is clean dissolved."

To regard the Countenance that shineth into our faces from on high is to be aware of the Countenance as shining from all sides into our faces, and from below as the foundation under our feet.

All is promised those who follow the high watch even a new heaven and a new earth.

Therefore, leaving all the shows of human misery ofttime and ofttime rejecting them, seek ye the Lord and His strength, seek His Face evermore.

'The Eighth was Simeon the hearkener." It is certain that the inner ear back of our outer ear does hear statements from every direction where the inner eye directs itself. The Siena Saint heard a voice from out the silence saying, "My daughter, think of Me, and I will think of thee." She knew it was Jesus the martyr, for the voice further informed her that he left her sometimes in order that she might feel herself deprived of consolation, and afflicted by pain.

How far this is removed from the voice of the High Deliverer speaking to Jeremiah, "They shall fight against thee; but they shall not prevail against thee; for I am with thee to deliver thee." Or from the heavenly voice speaking through the transfiguring Christ Jesus, "Nothing shall by any means hurt you"—"I am with you alway."

It is sign of ongoing in relation to that hidden objective toward which our inner all-compelling vision is oftenest set when we have inner auditions. And inner auditions from that direction we most certainly shall have. If a voice sounds on our inner ear that a boat is to be shipwrecked and we must not set sail upon it, we may surely infer that our inner vision has been ofttime turned toward danger of some kind, or menacings of catastrophe. If we hear a voice on the inner ear saying, "Fear not, I am God thy Saviour and I am thy people's Saviour"; we may know that our vision has been toward the High Deliverer, the Miracle-Working Almighty to Save.

The eighth lesson declares that ofttime glancing upward toward the Saving One we are saved, and as Paul was told, all they that sail with us shall be saved. Nothing shall have power to hurt them.

Is it any wonder that the wisely inspired ofttime have found themselves saying, "It is not so," when tales of danger and descriptions of calamity have been declared in their presence? "Nay, Nay," said the Brahmin high watcher.

"I sign it all away by the mystic activity of the cross I carry on my shoulder, symbol forever of blotting out, erasing, undoing the heavy burdens, taking off the heavy yokes," whispers the four-thousand year old voice of the miracle-working high watcher of forgotten old Egypt. The cross was once the symbol of such strong agreement with joyous Reality that all differentiations from joyous Reality were denied, rejected, blotted out.

Suffering, disappointment, poverty, degradation, death, those hieroglyphics against the Beautiful God, the Prince of Peace, the Miracle-Working Angel, are all erasable by denial, rejection, when they confront us as evidence of our past downward visionings. "If thou wouldest accurately put away all contentious words, O Child, thou shouldest find that truly the Soul dominates" —the Free Unspoilable Self makes manifest.

"Salute no man by the way." Salute no mighty claim except to wash it off the Ain Soph, Great Countenance of the Absolute, above thinking and above being.

All recognitions react. Notice the mysterious reactions that come from gazings toward a tormenting god: "A great desolation, the Lord increasing grief, pain that grows to such a degree of intensity that in spite of oneself one cries aloud," wrote the Carmelite Teresa, steady watcher toward a terrible presence.

High Science recognizes only the God who extendeth peace like a river, who, like as a mother comforteth so comforteth He, saying, "I am the Lord that healeth thee," "I give life to the faint, and to such as have no might I increase their strength."

It makes a mighty difference to mankind what description of God they look toward. The tormenting, angry, partiality-showing God of old Sainthood made them most miserable in mind, body and affairs. Their consciousness of their own wickedness was terrible. "Wheresoever thou findest self drop that self" wrote one of them. But he did not say, drop that imaginary God whose impress on thee makes thee dissatisfied with thyself.

"All the gods of the nations are idols. Sing unto the Lord a new song. Declare his glory among the heathen," shouted Ezra the learned descendant of Hilkiah the priest.

To endure as seeing Invisible Divinity is to extend all the faculties and sense the identity of all life with the Unlimited Supernal, the Free Self-Existent. To endure as seeing Invisible Divinity is to become one-pointed. "Only the one-pointed succeed," said the old Buddhist priests. When Sir Isaac Newton tried to explain the reason for his own great success he used words which meant that he had been one-pointed.

The seventh law of mysticism broaches the original heirship of all the sons of earth to One Divine All-Glorious Father. In the seventh we are told to keep our eye on the Lord Jehovah standing stately and majestic on the earth as the Self of our neighbor ever facing us, as Joshua saw the Lord Jehovah standing in the midst of moon-worshipping Jericho, and as King David foresaw the Lord always before his face. Face The God as Near and Awaiting.

This eighth law of mysticism urges standing by our recognition of the Lord Jehovah as the Self of our neighbor, in the face of all contentious appearances: "If thou wouldest accurately put away all contentious words, O child, thou shouldest find that truly the Soul (the free Self) dominates."

Every tongue everywhere seems to be proclaiming contentious words against the glorious Lord in the midst of man, against the Lord mighty to save, against the Angel of God's Presence, his own and his neighbor's Jehovah Sonhood. "I am unhappy," says the woman. Of what "I" is she speaking? Of the "I" self that we sense by seeing toward joyous Divinity, or toward tormenting imaginations? "I am losing my grip," groans the man. Of what "I am" is he talking? "I nearly perished," wails the backward-looker. Whose past is this historian recounting? Who has strength of silent rejection sufficient to "accurately put away" such dark descriptives of the Self of himself and the Self of his neighbor?

Have we not been told with the distinctness of trip hammers hitting bell rims that the True Self, the Son of the Highest is joyous, strong, imperishable? The "I" can say anything it chooses about itself. And all its sayings are vitalized by that subtler activity the inward visional sense, so that when the "I" chooses to say "I am unhappy," or "I am losing," or "I perish," we may know that according to law the "I" is blackhanding himself or herself with dark visions of himself that weave their words into cocoons of hiding, behind which only the vigorous at rejection can peer, enduring as seeing the Invisible, the Unspoilable.

There is the soundless Invisible of our neighbor, his strong Son of God. It is our true ministry to guide his speech into descriptions of his soundless Invisible. Isaiah speaks of the great rebound of such faithful ministry: "He that stoppeth his ears—and shutteth his eyes from seeing evil, shall dwell on high."

As evil is a weight, how can we help rising to happy heights as we reject evil? "No!" says the Sage of India. This is his way of putting away contentious words spoken against his neighbor's original Self. "I make the sign of the cross as if erasing marks on a tablet," signals the priest of Isis. And the gods of ancient Egypt were sculptured bearing crosses to signify that they stood for accurately putting away contentious descriptions of the glorious Invisible ever facing mankind.

"Be not deceived," said Jesus. This intimates that Jesus was brushing aside contentious descriptions and being one-pointed to the unspoilable Invisible. "Judge not according to the appearance," He said. This shows how persistent He was toward the Invisible as a waiting splendor.

Attention is the secret of success. Would Peace have folded all Firenze in its victorious arms, if Cosimo de Medici had not noticed it with his inward viewing, as a moveless Irresistible radiating forth from the non-resisting Antonino?

Would the hidden Lazarus have burst the bars of death if Jesus had not looked toward the One Life Undefeatable? Was not Cosimo ignoring the cries of hunger and fear in all his city, while the vision of Peace was holding him spellbound? Was not Jesus ignoring death while the vision of Triumphant Life held his gaze?

"Every tongue that shall arise against thee in judgment, thou shalt condemn," said the prophet Isaiah. To condemn is to declare useless. Is not a bridge declared useless when the authorities have condemned it?

When the attention is wholly engrossed with peace, or life, or strength, there is a tacit rejection of turmoil, death, weakness. When the attention is strongly distracted toward the cocoons woven by the dusky "I am's" of the downward watch toward death, disease, discord, we are told that we must use positive words of rejection to truly accompany our watch toward the Invisible.

"If any man will come after me (the Inviolate Unseen), let him deny himself," said the great Seer. Let him refuse the evidence of his outward senses, and watch, "What I say unto you I say unto all, Watch."

Their watch toward glory despite outward horrors has made the. mystics of all time the most powerful engines for miraculous good that the world has held. They have all been miracle workers. They have not forgotten that outward and objective signs always manifest when true inward viewing is married to true positive speech. "Tabitha, arise!" said Peter, addressing Unkillable Vitality, paying no attention to visible death. "Your son lives!" shouted St. Anthony of Padua, disregarding the messenger's death report. "Thou, Lord, holdest my precious gem! It cannot be lost!" said one facing the Owner of the spheres, and ignoring loss. "Thou *art the Saviour* of this woman from her own speech," said another stopping both ears from the hearing of evil. All these endured as seeing the Invisible Miracle-Worker, and for each a miracle was wrought. This is the Eternal Science. The ways of the stars may alter, the calculations of mathematics may readjust but recognition of the Almighty Uncontaminate shall not fail of the miracle of saving, healing, restoring, reviving.

"Therefore," says Micah, 700 B.C., "I will look unto the Lord; I will wait for the God of salvation: my God will hear me." "Therefore I looked to the healing Lord standing before me, and the dying baby lived," whispered the mystically trained nurse, 1916 A.D. This Science of the watch Godward shines brighter and brighter and works stronger and stronger as time wheels forward, finding more and more obedients who do the will to prove the doctrine. "Sing unto the Lord a new song. . . . Declare his glory."

Every invisible objective upon being acknowledged has its own *modus operandi;* its own steps toward manifestation; its own time of manifesting, and its own expressions upon arrival. "Therefore," said the son of Sirach, "do thou thy work betimes; he shall reward thee in his time." "Ye believe in God, believe also in (the) me," said the Greatest Expression.

As a right ministry we never forget that our neighbor has the Lord of his presence, his stately painless Soul, his Me, ever with him. We never forget that the Lord of our neighbor's presence is his rightful Self, which it is our ministry to bring forward. Our watch toward the High Deliverer wakes our awareness of the One Life ready to break forth in full shining as we face our neighbor. Therefore if we had not been told to reject man's false descriptions of himself we should find ourselves saying, "It is not so!" when we heard him telling of his losses or his pains.

"Remember only that He is looking toward you," said a certain Father Confessor to the Abbess of a convent. "Do not remember anything else. Exercise no self scrutiny." (See Résumé). So the Abbess remembered only that the Vast, Vast Shining Countenance looked toward her. She ignored her mind's reminders of her troubles. She ignored everything but the Shining Countenance. She was a true Obedient, for a stranger suddenly appeared before her offering her a bag of gold, though the stranger had no knowledge whatsoever that the sole trouble of the Abbess was lack of money to buy food for her waiting nuns. Thus was the obedient Abbess a plain demonstration of the promise, "Before they call, I will answer; and while they are yet speaking, I will hear." By her volitional absorption into Invisible Kindness she was visibly befriended.

"Holy Ann" of Canada spoke to Invisible Kindness on the subject of refreshment for the cattle, and suddenly a dry well was filled with water in time of great drought. (Holy Ann, p. 77.) A Jewish youth praised Invisible Goodness for making him a genius and much beloved. Genius began to wake in him and everybody began to love him.

A certain woman spoke praisefully to the Lord as Divine Beauty everywhere, and slowly beauty began to express in her own form and features. She praised Un-expressed Health, and bones and sinews, blood and nerves speedily began to express health. "Behold the beauty of the Lord," sang King David. "Oh that men would praise the Lord for his goodness, and for his wonderful works," sang Ezra, after keeping the same praises him-self he had urged others to keep.

We are told to ignore untoward appearances. We are told to *gather unto* with undivided choice. For what we *gather unto* we express.

The Hindus call the ever waiting Kindness, "Con-sciousness," or *"Brahma"* (from brih, to expand as Om-nipresence) ; and they show that every particle of exist-ence is a particle of Consciousness. Therefore we as consciousness gathering to Original Consciousness, ex-press its Beauty and its Health if we gather to Con-sciousness as Beauty or as Health. We express it as Strength if we gather to it as Strength. We express it as Triumph if we gather to it as Triumph. We express nothing of Beauty, Strength, Triumph, if we do not choose Brahma—Supernal Consciousness, in Its native expression as Beauty, Health, Triumph. We must keep on devoid of health, triumph, strength, if we as conscious-ness gather not to Divine Consciousness. This is the genesis of rejection of the devoid states called disaster, weakness, failure. "Loose thyself from the bands of thy neck, O captive daughter of Zion." To us there must be no God sending tribulation. Only the God letting the oppressed go free and taking off every yoke is our God.

There are reservoirs of Good back of all appearances of sickness or unhappy circumstances. "The eye of Judah is red with wine" to discover the reservoirs of Good. Judah means *praise*. He who praises the Unseen Providence just above his head experiences the Providence. He who chooses Providence puts away contentious appearances, by firm rejections, like the Abbess of Port Royal, or he ignores them like Cosimo de Medici, in the rapt vision of Peace. His eye, inspired to right viewing ("red with wine") , is the eye of wisdom that Solomon says "seeth precious things."

The *guru* sees intelligence in his *chela*. He does not stop to criticize the *chela's* stupidity. With steadfast gaze toward intelligence, like as Hufeland toward health, the *guru* enlarges the borders of intelligence. "Enlarge the place of thy tent," saith the voice of the Lord to Isaiah. "Salute no man (or claim) by the way," said Jesus. When the injured Chicago man talked to the strength back of his weak joints and muscles, he talked to the reservoir of good, the waiting Lord watching him. He saluted no claim of weakness hugging down over his bodily frame. And strength expressed itself so greatly that he who before could lift nothing, did suddenly lift heavy weights. "Ascribe ye strength unto God," sang David. And as we are always like the God we secretly describe, so David further sang, "God is the strength of my heart."

A boy in the cold is selling newspapers. Whose eye is "red with wine" to see the Lord strong and triumphant standing up tall and stately by or in the newsboy's presence, wilfully rejecting his rags? Whoever opens eyes red thus with inspiration to see the precious Life, soon sees the newsboy in some great University teaching the music of the spheres. The ways of the Lord are always miracles.

There are reservoirs of gold back of all gold pieces. On behalf of the facing Unlimited, erase, reject the limitations hieroglyphed over the gold pieces. This throws them out to the Universal Unlimited. Simon Magus had his "eye red with wine" to see the levitating principle back of the attraction of gravitation. "Lift me up," he said, ignoring all pullings earthward. And the Responsive Centrifugal lifted him into the air.

"I will make an everlasting covenant with you," saith the Lord. "Ask what ye will."—"And concerning the work of my hands command ye me." This Covenanting Almighty faces us everywhere. Even Talleyrand, foreign minister for Napoleon, found that strength, success, and haughtiness stood forth in his colleagues if he recognized them as strong, successful, haughty; and that they drooped in weakness, failure and dejection if he neglected them. Rachel the actress, a contemporary of Talleyrand, could see a thousand people smile, and they would smile, or see them weep, and they would weep. Napoleon himself could vision men as victorious or defeated, and they would execute his views of them. He wrote to his unfortunate brother Henry: "I have seen with pain that you represent everything to yourself on the black side. Take a resolution and stand by it with your whole strength." In other words, face something worth while.

When Marcion of the second century advocated the ascetic life, he did so because he saw that success lies in being one-pointed, and in rejecting all that distracts from One Victorious Objective.

Jesus asked mankind to regard him as the great expression *in toto* of the Victorious Unseen. Some said they must attend to their wives. Some said they must attend to their enterprises. Some said they must attend to their dead. They did not catch the mystic essential of His doctrine, that He that hath chosen Me, hath chosen the Finished Co-Efficient, touching all undertakings with flawless Finish, touching all undertakings with supernal completeness. For "He that hath seen me hath seen the Father." And all things are right with him.

The Parsees of old declared that ninety-nine persons out of every hundred die by reason of the evil eye. They see their own spot of pain or illness so distinctly that it expands into a sum total of shadow for them; and the sum total of shadow is the full manifest of devoid, or death. A certain commissariat in the federal army could always foretell what soldier was to be killed in the impending battle. He told this by the shadowy haze that he saw clinging to the soldier's face. After the war he could by the same token select which of his parishioners was about to *pass into devoid*. When this commissariat-clergyman learned of the principle of rejection, putting away, refusing death, disease, failure, on behalf of God the "o'er all victorious" then present, he spoke with great firmness to the shadowy haze whenever it claimed his attention: "Go away! I'll have nothing to do with you! Leave this man to his Life! O Life! Stand forth! Stand forth Free Life, bold and joyous!" And the mysterious devoid would give way to Life and Health, the rightful manifest of all mankind. Who can deny that this clergyman did perform the noble Christian Ministry, set into the form of command, "Heal the sick, raise the dead."

It is told of Pope Pius IX that his vision worked ill luck with unfortunate speed. Doubtless he was clairvoyant to troubles about to come to pass according to the law of cause and effect; and probably he knew nothing of the Trismegistian law, "If thou wouldest accurately put away contentious words, thou wouldest find free Spirit, untrammelled Life dominating." And the pope did certainly ignore John's beautifully executive treatment—"That thou mayest prosper and be in health, even as thy soul prospereth."

All people with the habit of looking for the best have the good and the favoring eye. The good and the favoring eye coalesces with the Deific eye: "We cried unto the Lord God of our Fathers" and He . . . "looked on our affliction, and our labour and on our oppression, and he brought us forth out of Egypt."

Joash saw himself as Lord undefeatable. Amaziah looked him in the face on the plains of Beth Shemesh hoping to defeat him. The lordly Self-viewing of Joash won the day for him as lordly Self-viewing always wins the day for any man. "Why criest thou out aloud? Is there no king in thee?"

Did not Hegel discover that "by oft recourse by inward viewing, the mind goes on to know and comprehend?" Mind is the ostensible and tangible nearest to ostensible and tangible outward activities. Hegel wrote his books about mind, because his inward viewing stayed not where his own words led him. Steadfastly maintaining his cause he would have written his books about the antecedent to mind, and so would have planted on earth the beautiful doctrine of viewing Godward, that the New Age might speedily usher in—

"When peace should over all the earth
  Its final splendors fling,
And the whole world ring back the song
  Which now the angels sing."

Pilate's wife urged him not to see Jesus crucified. But Pilate could not take his eye off the crucifixion. This premonitive viewing was suicidal for Pilate, as the vision of the crucifixion is always suicidal. "To the Christ that

never was crucified! To the Christ that never was buried!
To the Christ that never rose from the dead! To the
Eternal Almighty Christ, I commend you!" cried Elias
Hicks in one great great moment of ecstatic vision. To
endure as seeing the Triumphant Unkillable is to renew
the life forces and catch the breath of Omnipotence. But
Elias did not endure as seeing such Invisible. He alter-
cated with sinful adversaries, fellowshipping with Chrysostom's fateful forgetfulness.

Somebody must maintain the doctrine, that He that
beholdeth Me on high shall behold Me on all sides. He
that beholdeth Me on all sides shall behold that I am
watching him. Did not Hagar discover "Thou God seest
me?"

"The fourth beast whose look was more stout than
his fellows, made war against the saints of God, and prevailed against them, till the Ancient of Days came," wrote
Daniel, seeing past, present and future beastly strong
secret viewing toward devoids, as being some day defeated
by enduring vision toward the Eternal Almighty.

Steadfast vision *tangiblizes*. "There are eight conditions to right living, and the first is right view," whispered the mystics of old India. "With right glance and
right speech man superintendeth the universe," insisted
the far past Zoroastrian mystics. Francis d'Assisi sat with
closed eyes, inwardly beholding Jesus on the cross. With
profound sympathy of feeling he entered into the pains
of the wounded hands, feet and side of the image he thus
vividly visualized. Suddenly he himself showed wounded
hands and feet, and all the people cried, "Wonderful!
Wonderful!" Can you not see that his only wonderful-
ness consisted in his persistence of view? Where would the
world now be in demonstration if Francis d'Assisi had

persisted in beholding the Unkillable Almighty, the High and Lofty One inhabiting Eternity, "who healeth all our diseases, who redeemeth our life from destruction, who crowneth us with loving-kindness and tender mercy?"

Gregory the Great attested that St. Fridian changed the bed of a river by his strength of inward viewing, or one-pointed gaze. Dr. Gentry tells of a man who caught eczema from a picture of eczema. Hot focus of inward viewing *tangiblizes*. This is the genesis of the axiom, "He that findeth me findeth life."

"Turn ye even to me with fasting—and behold, the Lord will do great things." Fasting and circumcision, and the sign of the cross and sacrificing have ever been the outward ceremonies of mankind intended to signify putting aside the rigors of the law of cause and effect. Mankind everywhere vaguely senses his sinfulness, or darkly surmises himself in the wrong somehow, and therefore bound for punishment for the same. Why not, if his God is a punishing God and not the God who blotteth out transgressions?

Mankind has tried fasting from food and other normal goods in the hope of mayhap squaring his account with the great Lawgiver. Mankind has tried sacrificing his treasures of flocks and gold in the same hope. He has tried circumcision to set himself aside from human hordes, and seal himself to the God of his own description, still grovelling to placate the Lawgiver. Now and then one has boldly proclaimed that all these outward processes convey inward meanings, as that one should be circumcised from his heart's desires, or sacrifice himself for Principle, even to being burned at the stake or sawn asunder: "O ye stiffnecked and uncircumcised of heart and of ear," cried Stephen, "ye do always resist the Holy Ghost. As your fathers did, so do ye." Stephen is going far towards true circumcision, for the High Deliverer is always speaking wonderful words to him who endures as seeing the Invisible. How can the ears of the downward watcher, clogged with the stories of war and death, catch the voice of Him who saith, "I will instruct thee and teach thee?" "Walking in the comfort of the Holy Ghost."

Odilo of Cluny understood the sign of the cross as a blotting out and rejecting signal, and when he wet his fingers and made the sign of the cross, to indicate that he would have nothing to do with blindness chalk-marked across the shining God looking hitherward, blindness immediately disappeared from the eyes of the man born blind. By the same heavenly erasive signal he refused to see the ecclesiastic's great tumor; and it disappeared. The idiocy of a child was erased by this symbol of annulling, uttering forth from his inward viewing toward the Unspoilable Almighty looking straight into his eyes.

The priests of Plutonian Serapis made signs of the cross to signify their blackhanding with the evil on which their gaze was fastened. In this Christian day there are unwitting disciples of Serapis, first regarding the cross as sign of trouble, and second describing themselves on the devoid side. "I am all undone," says one. "I have been three days without food," says another. Both these are blackhanding their triumphing Lordship with devoids. And we who say, "I am sorry for you," are doubling their blackhand. We must silently stand by our true ministry: *"It is not so!"* we silently declare, offering our sacrificings of denial with secret shoutings, like the priests of Judah. We give our best praises to the Angel of their presence, their Jesus Christ Self, with face shining as the sun, and raiment white as the light. "Thus have ye done it unto the Mysterious Me," and prosperity and joy gleam boldly forth.

All subjective or secret images come to the ostensible. William Blake, the poet, in his subjective or inward viewings, revelled in the hells as essentials of God; but when he outwardly beheld the suffering children, he flamed with indignation. He could not bear these tangible ostensibles of his own hidden limnings. He forgot his doctrine that hells are essentials. We must have a doctrine that we are glad never to forget—the doctrine that Invisible Kindness comes out to heal the starving children, rewarding our undiverted, right secret viewings with love hurrying into childhood's daily life. "Is not this the fast that I have chosen? to loose the bands of wickedness, to undo the heavy burdens, to let the oppressed go free, and that ye break every yoke." Fast ye from stories of pain, disappointment, poverty. They are only hieroglyphics painted across the joyous Presence. They represent looking away from the true God.

There is a force more mighty than mind, more potent than thought. It is the "Dayspring from on high" that falls down over the upward watch, giving light in darkness, and guiding into peace. It is the resistless Holy Ghost, waiting to be heard by the ears from the hearing of evil. Micah heard it telling the Jews that the Lord was their redeemer from the hand of their enemies. Zephaniah heard it telling them that the Lord in the midst of them should cast away their enemies so that they should not see evil any more. Habakkuk heard it telling him to write his vision and make it plain.

Euripides wrote out descriptions of his visions of outwardly unseen men and women, gods and goddesses more powerful in battle and more daring and original in social encounter than any his outer eyes had beheld. So steadfast was his gaze toward these transcending images, that they made him know the words appropriate to their state and greatness. Then when Euripides the son of an herb seller, entered the theater, the Athenians rose as when their king appeared, and they cried that "the glory of the Athenian stage descended into the tomb," when Euripides ceased writing his immortal plays.

Whoever is one-pointed to any objective begins to know its speech. Thus the *eighth* stone of character is the beryl stone, significant of one who hearkens to the teachings that wait at the silent heights. It stands for writing out the teachings. "What thou seest, write and send it unto the churches" said the voice of the Lord to John the Revelator. "Write thee all the words that I have spoken unto thee," is said to the man that heareth. "The Eighth was Simeon the hearkener."

Only one objective is worth our one-pointed, undivided attention. Only one objective has healing on its returning beam. That objective is the Watcher hitherward, whose soundless call is, "Look unto Me"—"Behold I bring thee health and cure."

Archimedes gave his undivided attention to mathematics. *"Noli turbare circulos meos,"* he said to the soldiers driving their swords toward him. Could any attention be more undivided? But his beautiful mathematics did not save his life. For mathematics has never said, "I will contend with him that contendeth with thee." Beethoven with undivided attention to notes that burst forth into delectable sounds, was hard of hearing, for music has never agreed to unstop deaf ears, open the eyes of the blind and cause the lame to leap like harts. Only the Healing God has promised such lovely service. A deaf child prayed to God, and her ears were unstopped. A blind women prayed to Him, and her eyes were opened. "Therefore I will lift up mine eyes unto Thee, O Thou that dwellest in the heavens." "He bindeth up their wounds."

We are all walled into chambers of imagery, taught Ezekiel. Of course! Why not, if our executive sense is drawing pictures from other objectives than the Free Universal? Why not, if our attention is drawing ideas from minds in all directions set toward us with their knowledges all foolishness to Original Wisdom? How unhappy was the little Dalai Lama of Lhasa, always reflecting the minds of those who came near him, speaking the almost forgotten dialects of strangers, and reading their innermost thoughts! How could he help even dying young under the burden of so much imagery?

The greatest adept of time was Jesus of Nazareth, who drew his knowledges altogether from God. "What went ye out into the wilderness to see?" He asked, "a reed shaken with the wind?" A being standing proud and glad at beholding your favorable opinions; drooping and ashamed at your disapprovals? Whoever stood so on his own feet, independent forever of the estimates of mankind! He ignored their estimates. What he was he was, and they could not kill him with secret hatred or with open castigation. "No man taketh my life from me— I lay it down of myself," he said. "What went ye out for to see? A prophet? yea, I say unto you, and more than a prophet." Samuel had been a great prophet. He was one who like Talleyrand, and Rachel, and Napoleon, laid his views on mankind so thickly they could not hold their own. "Thou art a victorious soldier," laid Samuel's estimate on young Saul. And Saul rose up a victorious soldier. "Thou art a king," pictured Samuel, and up went Saul to the throne of Israel. "Thou art no king," imaged the same prophet, and down went Saul, mindless and throneless.

But Jesus never offered to paint men any different from their original estate. "Call no man your father upon the earth. One is your Father." "Ye are the light of the world." "Not of the world, even as I am not of the world." He put away from his Apostles the estimates of the whole world and their own estimates also. They were dull fishermen in the world's esteem, but he wrought forth their original light as the noonday, and their native radiance that height nor depth nor things present nor things to come can ever dim.

There is one Shining Me, the Unalterable Fundamental, the High Reality of every son of man. This is the handwriting of God. All other sights and sounds of him are fictitious handwritings, as erasable by the right process as chalk marks on a blackboard.

It is the province of this *eighth* lesson in heavenly law to woo mankind to willing erasure of all fictitious estimates. "Man can never behold the face of the Father except in the *eighth*"—was an inspired utterance of Saint-Martin *"le philosophe inconnu."* This was Saint-Martin's way of putting Isaiah's words, "He that is escaped shall come unto thee to cause thee to hear." One is escaped, or free to the view who has no imagery hieroglyphed across his presence. "Nay, not that!" says the Hindu, to all descriptions of the Lord of Life and Spirit.

"He is not Life, but Cause that Life is. He is not Spirit, but Cause that Spirit is. No descriptive fits the High Cause," wrote the Egyptian wise men as they laid the symbol of high denial on the shoulders of their sculptured gods.

"Blotting out the handwriting of ordinances against us—took it out of the way, nailing it to (signing it by) his cross," wrote Paul, learned in the significance of all symbols. "Beware lest any man spoil you through philosophy and vain deceit, after the traditions of men, after the rudiments of the world."

"The accuser of our brethren is cast down," said the voice of the angel to John the Revelator. Accusation does fall down. It is the tongue silently or audibly detailing descriptives that do not apply to the Actual. "Then the ruffian looked at me, and wrought against me strange diseases," said the Parsee, speaking for mankind become self-accused, or neighbor-accused by low watching. "So mayest thou heal me thou glorious Manthra Spenta." *Manthra Spenta* is Lord, and lordly recognition of what lies back of the ruffianly accused heals. It is the Parsee way of urging the healing *high watch*.

Even objects of Nature, trees, plants, winds, waters, have been ruffianly accused of hurting powers they hold not in themselves. "The waters drown, the winds devastate, the plants poison," we are told. Not in themselves. Have they not all been pronounced good from the beginning? Did the waters drown Peter while his eye was on the Lord mighty to save? Did the winds overturn the ship while the sailors looked to the Master's face? Can tea make the Angel of God's Presence nervous? Or coffee cause Divine Intelligence to degenerate? Can rum spoil the beauty of God? It is ruffianly to speak of

man or item of Nature as having the power to hurt.
Look at them as glowing with the face of One ever watching from every infinitesimal point of the universe.
"Wheresoever thou lookest there I am." Take off the
world's estimates from that face, and lo, Beneficence only!

Do not go back on the knowledge that "Thou God
seest me" from every direction. Thus are we dead to sin
or aberrated viewings. Thus are we dead with the un-
condemning Christ, that we may live with Him. Thus
are we *tabula rosa* for truth from the circumambient
kingdom in which we truly dwell.

"*Multum incola*," said the great Bacon. He was
sensitive to world conditions. They made him ill and
dulled his intelligence. He did not care what men called
him so they kept out of his way and left him *much alone*
to be sensitive to the wisdoms that fill the ethers. Thus
being washed, set free from puny knowledges, he was
taught new principles, which made him to be declared
"wisest and brightest" of his generation.

Some way of getting *multum incola*, or much alone,
has been the effort of all who would be powerful with
Truth. Do not the Brahmins reject outward sounds and
sights, and inward memories of sorrow and human vexa-
tions, until they are left alone with Brahma, or Universal
Consciousness and its mysteries? Have the Brahmins not
learned, by making themselves thus *multum incola*, many
truths of the Universal with which they have practiced
till their powers have astonished the world?

"Become relaxed," say certain practitioners among
us. "Drop sights and sounds; drop thoughts; drop emotions; become blank for a few seconds; then allow one
positive thought to take possession. Hold the thought
firmly. It will demonstrate in time in outward condi-

tions." This is their way of letting go, rejecting, on behalf of a thought. There is One above thought, above being, wrote the Seers of old Cabala philosophy. "It is The Ain Soph," they said. "Take no thought," said Jesus. "Look up to the fields white for the harvest."

Paul speaks of dropping the traditions and philosophies of men, that the life of Christ may take possession of us. Does not the magnet drop all sticks and stones and dust that its own may gather unto it? See the needles and the nails and the steel filings that the magnet loves, hurrying to gather to it!

> "So runs the good with equal law
> Unto the Soul of pure delight."

"Thy lot or portion in life is seeking after thee," said the Caliph Ali.

"Through the voice of the Lord shall the Assyrian be beaten down," wrote the prophet. "The Assyrian" is the power of darkness.

"The Messiah cometh to walk with men, when they hearken to the voice of the Lord," promises the Talmud.

How can men hearken to the voice of the Lord with its soundless wordings, if the sounded wordings of their neighbors are thick on their ears? How can we know the grand verity of our neighbor's Actual if we accept the outward shows and blatant estimates that make him cower and cringe to old age and death, disease and poverty?

This is a trumpet call that we take today, to accurately put away all contentious estimates and relate ourselves to the Right Estimate. It is the call of the Eight: "In the eighth month . . . came the word of the Lord . . . to

Zechariah, . . . the Son of Iddo the prophet, saying, . . . Therefore say thou unto them, Thus saith the Lord of hosts, Turn ye unto me, saith the Lord of hosts and I will turn unto you, saith the Lord of hosts."

How far reaching is the injunction, "Go, wash in the pool of Siloam"—the curative waters of *erasure*.

Go, gather unto the Unlimited. Erase age, destiny, inadequacy, disadvantage. On behalf of Divine Presence, One and Uncontaminate, reject its unlikeness.

Stand alone with Almighty Verity. Be one-pointed to the Ever-Facing Lord Jehovah. Declare the Unburdening Free God. Be undimmed heralding Star of Messiah's Bright Morning.

IX

MINISTRY

PROLOGUE

The Italian Flagellants of the 13th century lashed themselves and gashed themselves with thongs and scissors but they felt no pain for the zeal which arose from some central flame within them permeated them through all their flesh and of itself neutralized flesh sensations. Even their minds were so alive with the central flame that they sprang the bounds of self pity in unconsciousness of personal hurts.

Some people explained that it was innate Will rising to new action with a new sense, as "I willed and new sense was given me."

The persecuted English Quakers of the 17th century declared that no blows or stones which their enemies administered upon them with ferocious energy were felt by them, for a central fire rose up in them neutralizing outward sensations by its ecstatic flame. They called it God willing in them.

Martyrs of old sang hymns in the midst of their enemies' hot bonfires, and their faces were seen to shine with inward ecstacy even as Stephen's face shone while the hurling stones were cutting him. They "saw his face as it had been the face of an angel." (Acts 6:15). The martyrs, like the Quakers, called the mystery of supernal sensation, "God willing them to will and to do."

Some people have called that conquering central flame which can so overpower all human sensation and mental awareness *The I AM* of man. Some have called it *The Spirit of man.* This was Solomon's name for it. Some have called it *The Spirit of God.* Ezra called it *The Spirit of God.* So also did Ezekiel the prophet who was named *The Strength of Jehovah.*

Jesus, the All-overshining Jesus of Nazareth, declared, "The Spirit of God is upon me." He also said "I AM," and transcended all other martyrs in the risen splendor of the central flame common to but hidden deep in all mankind. He knew the risen vigor as *Will.* "Whosoever shall do the will of my Father which is in heaven the same is my brother and sister, and mother." He shouted of His glorification despite the wounds of the cross, and cried aloud of His almighty identification with Universal Spirit, "Into Thy hands I commend my Spirit."

Socrates hundreds of years before the time of Jesus felt himself outshining himself in the face of threatening martyrdom, and said to the judges, "I go about trying to persuade both young and old, not to busy themselves about their bodies or about money more than about their soul." "Manifestly it rules in us, a King."

It is no wonder then that over the cross of Him who picked Himself up out of martyrdom and showed Himself Indestructible Soul, they wrote, "Jesus of Nazareth, the King of (the Soldiers of God, the Israelites) the Jews."

One historic life of this martyr-transcending King, lately written, lays such stress on the pains and torments of his martyrdom that people shudder and weep when they read it. And the *supposedly* greatest praise that has been lavished on an eloquent New York preacher reads: "We can almost feel the blood drip on our heads as we hear him preach the Cross."

But the true printed Life of Jesus should glory in telling of his sense of outglowing all wounds on a still grander scale than the Flagellants and Quakers and noblest of earth's martyrs, because He knew himself with ecstatic zeal as *God Himself* proving the rightful status of all mankind. "Thou being a man makest thyself God," they said.

Mystical Science which shows how the Secret Flame can be wooed to rise superior to all human laws physical and mental, tells of an Over Splendor of attention toward us ever wooing our attention till its likeness in us rises into Oneness with its Splendor as Undefeatable Greatness—the greatness of Unconquerable Will.

"By so many roots as the marsh grass sends in the sod,
I will heartily lay me ahold of the greatness of God."

"He sent from above, he took me, he drew me out of many waters"—even the waters of humiliation.

It is not a matter of consciousness or unconsciousness of mind or body. It is not a matter of right or wrong of human calculation. It is the New Kingdom's Super Life.

This ninth lesson teaches to stand by our Super "I" with its high language, till its language fruitings lay out before us our true life field. It teaches to find that hidden Self! It teaches to speak from that! It teaches to choose what we truly will, and shows it as Universal Will as it was in the beginning, is now, and ever shall be, out of the reach of death and those shadows of death, the hurts of flesh and of mind.

E. C. H.

IX

MINISTRY

Numbers were alive to the ancients. Notice Parmenides facing The One till form and differentiations disappeared, the cosmic order let go and he was alone as if he were The Alone. There is but One SELF and Thou art That Self.

Five was a living number to the Hebrews. It signified The Representative, as Joseph for Pharaoh, as Jesus for Jehovah. Its symbol was the sardonyx, which held the engraving of the royal "I Depute."

Apollonius of Tyana would not allow his disciples to speak the number *nine* aloud, as it was to him the magician's number—the number of enchantment. If sounded it was likely to hit the Super "I" having its location in a point back of the breast. To hit that "I," or speak forth from that "I," would be like wishing the words on somebody or something. Learn to wake the hidden "I," but note the kindling mystery of it as the living black coal lingering in the breast waiting the bellows breath of right declaration.

To mankind the old "I" must aver according to appearance, as "I am sick," or, "I am overworked." This is *its* gospel truth. But the "I," as kindling coal, avers, "I living vigor! I godly strength! I decree!"

Ye shall eat of the old fruit—ye shall eat of the new fruit, according to kindling or non-kindling of the words by recognition of the "I" Spark. "My spirit shall not forever be humbled in man"—"I am the light. He that eateth me shall live by me."

"I" is a fodder. Man shall eat of the old fruit till the *ninth,* or till willing to take up the Super "I's" great sayings, the "I decree!" At the *ninth,* man is willing to refuse the old fruitings. He declares as Christ in him, "I the Alive, I the Eternal, I the transfiguring and transforming all things by my 'I decree'."

The ancients cherished the scarab that drops its seed into a dirt ball for universal atmosphere to stir it to break its covering and fly as winged creature. They named it symbol of the resurrection of man as determining God Will from the encasement of shadow body formed of matter woven with mind threads, the stubborn web of hiding, "the shadow system gathered round the Me."

> "Rise my soul, and stretch thy wings,
> Thy better portion trace."

Nine is the Assumption, the Joy of Mary, the taking for granted, the supposing of a thing without visible proof, the unwarrantable claim. The Great Assumption is the Great "I" Spark speaking its own joyous facts, which come forth glorifying the outer man with new fruitings.

Nine stands for the closing of an old cycle and the opening of a new cycle, the old touching the new and the new touching the old like beads on a rosary. The great mystics have prophesied that the Messiah appears at the beginning of some nine cycle as the old "I" giving way to the new "I." The old "I" telling the truth of itself and others according to appearances, and the new "I," or the awakening "I," telling truth according to the kingdom of the Super Spark, till it flashes its dominion over

animate and inanimate. It is the daring response of the
Me, the "**I**," praised, according to the fifth revelation,
or the Fifth study.

It is indeed the swift discovery with **Hugo of St.
Victor** that the Highest God and the Inmost God is one
God.

Acknowledge the Highest, the Great Countenance of
the Absolute, and its "I" Seed awakes with the Assumption, the Unwarrantable Claim, the Word of the Lordship "very nigh unto thee, in thy mouth and in thy
heart, that thou mayest do it," as Moses said to the
children of Israel in the land of Moab. "I am as strong
this day, as I was in the day that Moses sent me (into
Eshcol of Canaan). And now, Lo, I am this day four
score and five years old," declared Caleb son of Jephun-
neh, who stood with good report from Eshcol as easily
manageable by Israel, when others said there were unmanageable giants to meet.

The divine "I" that has its location in a point back
of the breast, whose flame Apollonius asked his disciples
to stir with good reports, is still alive in every breast on
earth, and ofttime assumption of the things It sees, It
feels, It knows, will cause them to show forth, brightening all the paths in which we daily move with the un-
precedented, the unbelievable, the unexpected.

"Where is thy spark, Lanoo? Speak thou from that!"

The ministry of acknowledgment and recognition
into which we are hurried, is wide and far reaching, or
near and intimate in demonstration. according as our
vision proceeds far or keeps close. "I will fetch my knowledge from afar," said young Elihu. "Remember that thou
magnify His work which men may behold."

The forty Christians enrolled in the twelfth Roman

legion lifted up their hands to God in high heaven, when the Quadi seemed about to destroy the legion. And Marcus Aurelius Antoninus wrote it down for all ages to notice, that a Roman victory was wrested from the Quadi by the God of the forty Christians, and not by the swords and spears of the vaunted legion.

The original Francis Schlatter cured thousands of people by sensing that his God was near at hand, close and compassing, with only just one kind and merciful streak in His mighty composition, and that it was His will to heal through the willing Francis Schlatter. His God would not let him be fed, or warmly clothed, or rested when weary, but He would touch Schlatter's hands with curative balms for his neighbors.

The science of the recognition of the Presence of God in the Universe, by the uplift of the vision as to divine Beneficence, is the oldest science maintaining its stately march through the long non-acceptance of the human race. Abram, 1920 B. C. is told to lift up his eyes. Dante, 1265 A. D., declared that blessedness more depends upon vision than upon loving, for loving follows on after vision. It is told of a certain Bishop that his lofty character and great attainments were due to his ofttime gaze toward the glory of the Highest. We also may show forth lofty character and great attainments by ofttime gaze toward the Highest, the great Countenance of the Absolute, ever facing us.

"The Deity onlooketh thee, onlook thou the Deity."

The one mystery of man's accomplishing executiveness is that whatever he persistently recognizes and acknowledges surely comes forward and deals with him. Recognition and acknowledgment are the *"two-law"*

from which something must come forth, worth while
or not worth while. Man must woo and woo the Majesty
and Responsive Beneficence ever facing him. His wooing
will be astonishingly rewarded. "Go away," said the elder
Gounod to his son; "burn your composition, the Muses
have not called you." What did the elder Gounod know
about wooing the Muses till they gladly come with their
music, or their statuary, or their architecture, such as
this present world has not set eyes upon? Even the Moekel
terrier disputes Gounod's position that the Muses neglect
some people. The little dog did prove per contra that it
is some people who neglect the Muses. By much wooing
of the dog Rolf's intelligence Mrs. Moekel caused intelligence to speak forth through him with amazing answers
to mathematical and philosophical questions.

What wonderful music was closed down upon from
expressing through young Gounod, when his father forbade his wooing the Muses! What enchanting singers
have been silenced by teachers who did not help enlarge
the smile of the song-muse the young voice was trying to
woo—Voice that might have stayed the wars of empires!

"Muses" were ancient names for the Universal All-
Power, demonstrating according to recognition and
wooing.

Our eighth lesson urged recognition of the High
and Lofty One as All-Power, and as the Universal,
Rounded, Super Worth-While, into whose glorious
oblivion all limitations and false estimates may be
thrown, by casting them away, having nothing to do with
them. "*It is not so!*" is the eighth lesson's masterful shout
to every word that disputes all Judah praise of the everywhere watching Amen Ra, or Brahma, or God. Even

the values of man as good in contradistinction to bad are to be circumcised, sacrificed, rejected, on behalf of the High First Estate above the pairs of opposites good and bad, life and death, pleasure and pain. We must throw everything out to the ever present First Estate of the Universally Desirable.

This ninth lesson calls attention to the Undifferentiated, Ever-facing Original, which the Egyptian priests called Soul, the Magi called Self, the Hebrews sometimes called God and sometimes called Angel of the Presence, and which truly is Free Spirit, Free Beneficence, Free Zeus, Free Giver. It is that Dominant of which the ancients were speaking when they promised that whosoever should put away all words contending against high praises, should find Free Spirit, Immortal Self, dominating.

Great promises have ever associated with the ministry of rejections, denials, sacrifices, circumcisions:

"Grace, mercy and peace from God . . . teach no other doctrine . . . neither give heed to fables . . . that thou mightest war a good warfare," wrote Paul to Timothy, the boy Bishop of Ephesus. "For God hath not given us the spirit of fear, but of power, and of love, and of a sound mind—circumcision made without hands—putting off the body of the sins of the flesh by the circumcision of Christ —quickened together with him, having spoiled principalities and powers."

Did not Bhaskaranand of Benares find himself a powerful healer, after rejecting the influences of the world—"triumphing over them in himself," as Paul noted of Christ?

Coming thus to our own free Self, our own divinity

Spark, our own "I decree," clarifies the atmospheres of false instructions. Heretofore hidden truth comes boldly forth. I the mighty Truth. I the awakener to life eternal. "The Highest God and the Inmost God is one God."

"Thy words were found, and I did eat them, and thy word was unto me the joy and rejoicing of my heart." So wrote Jeremiah in a moment of freedom from his own contentious complainings. He experienced the outward joys of Victorious Assumption, the living *nine* of Apollonius.

There are promises with hearkening to the Heights: "Whoso hearkeneth unto me shall dwell safely, and shall be quiet from fear of evil"—"A friend of the bridegroom which standeth and heareth him, rejoiceth greatly"—"I will give you pastors according to mine heart, which shall feed you with knowledge and understanding."

Thus we are told how satisfying is "I" Truth! We always feed on the Super "I's" great claims. It affects our constitution much or little. As an eyelash falling into a pool affects it little, but a meteor falling in affects it greatly, so we are stirred greatly or slightly by the truths we hear and proclaim. Does it stir your life pool to be told that God is great and we know him not, as spoken through Elihu the Buzite? But if you should yourself hear one of the soundless truths that the great Ain Soph holds hidden in the etheric silence round about your head, especially waiting for your own ears only, promising that the desire of your heart is fulfilled this day, you would tremble from head to foot with joy. You would be found declaring, "Thy word in me is joy to mine own heart. I Joy, ray forth joy. I ray forth my own I Am, I Will, I Can."

Joy that inwardly stirs by truth from on high is a
healing energy, or Saviour, or Joshua. Joshua, or Joy,
particularizes the spot or the scene of its triumphing
"I decree." Joshua stretches forth the spear in his hand
toward Ai, the mass or heap of spoiling principles as the
"I's" old fruitings, according to its gospel truth speaking as things appear, that have gotten the better of man
as sickness, sorrow, or despair; and Joshua, aware of the
"I am Captain of the host of the Lord" facing him, taketh
not back the hand wherewith he stretcheth out the spear,
till all the inhabitants of Ai are utterly destroyed; or the
spoiling principles annihilated; and in the precinct is
no city save Beth-el, House of the Lord Uncontaminate.

In ancient times the spear, the stick, the staff, the
sword, the sceptre, were used to knight the commoner,
or to transform the sick man into the sound man, the
timid man into the bold man, the common man into the
titled peer. Elisha's staff was often used with this transforming effect. Kings to this day strike with sword or
sceptre the shoulders of certain subjects of their realm;
and Mr. Henry becomes Sir Henry, and my friend becomes Your lordship. "Take thee one stick and write . . .
Judah (the unseen eternal) . . . then take another and
write . . . Joseph (the manifest external) . . . and they
shall become one in thine hand," said the voice of the
Lord to Ezekiel the most mystical minister among the
ancient prophets.

Jesus did not take a visible stick or sceptre to touch
the beautiful Judah Self of the Joseph self of the corpse
body of Nain, to make the Judah beauty glow and gleam
and glorify through rosy lips and sparkling eyes and
stirring blood as living widow's son at the city gates. He
laid His hand on the Joseph self while steadfastly looking

toward the ever-waiting Angel of the Presence, flying swiftly, mighty in strength, ministering Spirit, doing wondrously; and He told His disciples to thus lay hands on the sick that they might recover their seemingly lost immortal Judah Self, the ever-near Angel of the Presence. He drew the waiting glow and glory of man's Angel of Life ever present, to gleam and gladden him as living strength, even when he seemed dead. "I drew them with cords . . . with bands of love and I was to them as they that take off the yoke on their jaws."

"If I keep looking to a gazelle leaping from rock to rock, I dance the gazelle dance so that people clap their hands," says the child dancer of the New Age, harking back to the days of Aristoxanus, who taught to dance according to soul promptings and inward visioning.

"Keep your eye on the Eternal, and your intellect will grow." "Honor and fortune exist for him who remembers that he is in the presence of the High Cause." Therefore, "Unto thee, O my Strength, will I sing, animating my particular Self with Thy Universal Energy! O Thou I, and I Thou!"

"Glance up often," writes Conan Doyle "to the spirits of the dead." Hosea the prophet would say that the great writer is calling attention high, but not to the Most High. Practice of the Presence of the Most High, the Great Countenance of the Absolute, the Ain Soph, caused Nahum the prophet to cry out, "Watch the way, make thy loins strong, fortify thy powers mightily!" For the loins do show strength by reason of high watch, and the "I Am Strength" is our giant truth.

"And . . . when he was in a certain city, behold a man full of leprosy (spoiling principles), who seeing Jesus fell on his face, and besought him, saying, Lord, if thou wilt, thou canst make me clean. And he put forth his hand, and touched him, saying, I will; be thou clean. And immediately the leprosy departed from him."

It makes a difference who is reporting a miracle, as to whether the misery in its departure is mentioned, or the joyous state that remains is proclaimed. Luke the physician when reporting a case lays stress on the malady and its exorcise, as in the case of the "man full of leprosy," which "leprosy departed." So Joshua lays stress on the destruction of the spoiling principles massed as the old "I's" gospel truth.

Mark, the founder of the Alexandrine Church, writes with burning brevity of the happy estate of unmolested cure: "And he took the blind man by the hand . . . and when he had spit on his eyes and put his hands upon him, he asked him if he saw aught . . . After that he put his hands again upon his eyes . . . and he was restored and saw every man clearly." Mark does not say the blindness departed.

Again, notice the unction with which the physician Luke details the distressing condition of the synagogue worshipper: "And, behold, there was a woman which had a spirit of infirmity eighteen years, and was bowed together, and could in no wise lift up herself. And when Jesus saw her, he called her to him, and said unto her, Woman, thou art loosed from thine infirmity. And he laid his hands upon her; and immediately she was made straight and glorified God." Notice how he threw the spoiling fruitings of the old "I" away into Universal Solvent!

So efficacious in drawing forth to visibility the Stainless Spark, or Judah Self of man, by the laying on of hands, was the Apostolic succession in Christian ministry, that for more than fifteen centuries the Christian minister laid with sacred tenderness his ordained hand upon the Joseph breast of his sick parishioner, whoever he might be, and repeated the accepted Christian formula for making the invisible Judah effulgence one with the outward body: "As with this visible oil thy body outwardly is anointed, so our heavenly Father, Almighty God, grant of His infinite goodness that thy soul, inwardly, may be anointed with the Holy Ghost, who is the Spirit of all strength, comfort, relief and gladness. And vouchsafe for His great mercy to restore unto thee thy bodily health, and strength, to serve Him, and send thee release from all thy pains, troubles and diseases, both in body and mind—through Jesus Christ our Lord." And wonderful were the restorations of the Lord Self to the man or woman who had lost sense of vital oneness of Judah and Joseph in their own bodily form.

"Labor to keep alive in your breast the spark of celestial fire, the quenchless flame," said some of the mystics. Remember the smokeless fire, the effulgent centre capable of *infinite procession in extenso,* of far extending radiance—"The Stainless Spark," of which Dante writes.

When Jesus touches the leper and says, "I will. Be thou clean," He is surely feeling the sweet fires running along from the God Self of Himself through the palms of His hands and the tips of His fingers, as He sights with mystic energy the unseen Angel to wake joyous soundness where no waking seemed possible. "I drew them with the bands of love, and they knew not that I healed them."

"Take thee one stick and write . . . Judah, then take another stick and write . . . Joseph . . . and they shall be one in thine hand."

The Lord mighty to save stands stately and majestic everywhere facing us. "Two are ever in the field. One shall be taken, the other left." Call, "Stand forth, life! health! strength! Come ye blessed of my Father! It is God's will that you stand forth strong, and glad, and free! It is my will that you stand forth strong, and glad, and free! It is your will that you stand forth strong, and glad, and free!" This memra, or *Will Word,* is the one harmonic chord stretching its soundless note, world without end, through men and angels, from the just, the mighty Will filling the universe.

Only he who has felt the blaze in his own breast can sense to the full the vital fire that burns in the breast of his neighbor, when his form seems lifeless, unconscious, past knighting with his own divinity. Therefore it is that all lessons of Ineffable Mysticism return again and again to the relation of one's own Lord Self to the High I AM, the Lord I AM ever standing before us. "I go to the king," said Esther. To all previous outer experience it was death to go to the king; but Esther felt the celestial "I Go," in her breast, and faced the Invisible Lord standing masterful and majestic where other people saw only Xerxes the terrible.

"Old born drunk" who had never had any will or mind, laid his hands on his breast, where the Spark Celestial glows in all men, and said, "I am determined to find God." And all the criminals and debauched of his part of London turned out with heads bowed and hats in hand, because "Old born drunk," blear-eyed, will-less sot, found God!

"I praise God that I am healed!" shouted one who felt his Spark Celestial stir to speak as truth speaks by inner Lord despite outer diseased contentions. And though all the people sided with his outer contention, or diseased form, he spoke again more loudly still of the Lord Spark, burning truth in his breast, "Praise God! He has made me whole! I am healed!" And his truth of his own Lord Self transfused and uplifted and out-shone so that his outer frame glowed with the beauty of health, and all those who had sided with his previous showings were confounded. "Let them be confounded that are adversaries to my soul," cried King David, sensing the glow of the Christ Spark in his bosom, the meeting place of the majestic Lord with every human form.

"I cannot retract—I am That!" exclaimed a young girl who had felt the glow of the Stainless Spark in her breast. And though she was held as untruthful, she fulfilled in herself all that she had said of herself "I am That!" She had spoken of the future as in the present tense, and that alone was her offense. Let us make no blunder here: The verities are in all tenses the same. The "I am That" belongs to us now as it was in the beginning, and always shall be, unabrogated forever. Whatever we will be we are, as Lord Self, ready to kindle the demonstration of "I am That," according to our acknowledgment. "Take thee one stick and write . . . Judah, then take another stick and write . . . Joseph . . . and they shall become one in thine hand," is our privilege. For the great things of our Judah Self are the things our Joseph self can manifest.

This has been called knowing the Self, or recognition of the pure Splendor inherent in all mankind as Omniscience, Omnipotence; Self Spark fearlessly waiting till *I Am That* be boldly spoken.

"Where is thy Spark, Lanoo! Speak thou from That."

Joy follows hearkening to the voice of the ever-speaking Highest. "Thy word was unto me the joy and rejoicing of mine heart." And joy is strength: "The joy of the Lord is your strength." And strength is the bread of heaven: "Man shall feed in the strength of the Lord." And joy in the Lord as strength is a healing radiance. "The merry heart doeth good like a medicine." "The Lord loveth a cheerful giver," or a giver of cheer. For they "tread on serpents." That is, they bring forth health and strength and wise words from their neighbors as the sun draws forth the serpents of the stony fields. Serpents were in old days symbolic of health, strength, wisdom.

"Canst thou draw out leviathan?" asks the great Voice addressing Job. "Leviathan" is the strongest beast of the sea. The strongest constituent of man's constitution is his joy chord. Canst thou make thy neighbor joyous? He is surely cured of all his maladies if thou canst wake the joy that slumbers unstirred in his being. Jeremiah had three diseases from grieving: "Mine eyes do fail with tears—my bowels are troubled—my liver is poured out." No heart was ever stirred to healing joyousness in Jeremiah's atmosphere. But a man in the Himalaya mountains felt the smile of God so ardently, that people went from miles around to get kindled into health by the mysterious influence of the far extending radio-activity of his smile.

No condemnation abides with the awakened joy spark. "I came not to condemn," said the healing Jesus. "He that is of a merry heart hath a continual feast," said Solomon. "His barrel of meal," his barrel of healing power, "stayeth not." As the lily is not lessened in fragrance though a thousand people smell thereof, so the bottomless pit of the God Spark stayeth not, is not lessened, by much *procession in extenso.* "There is that scattereth and yet increaseth" is true of the *I am That* of myself, and of the *Thou art That* which I ardently proclaim to my neighbor. There is no limit to his joy fountain, his "I" with its divine fruitings.

"Go thy way," respondeth the Mighty Lord of my neighbor's presence; "eat thy bread with joy, and drink thy wine with a merry heart; for God now accepteth thy works." "Ye shall be named the Priests of the Lord. Men shall call you the Ministers of our God." "I will make thee a joy of many generations."

This *Ninth* tells of the joy that comes by communing with one stronger than Nature's forces and wiser than man's mind. "And your joy no man taketh away." "For he that communeth with me strengtheneth." Gideon communed with the Lord God Invisible. He endureth year after year as seeing Him. "And the Lord looked on Gideon, and said, Go in this thy might, and thou shalt save Israel from the hand of the Midianites . . . and the Amalekites, and all the children of the east lying along in the valley like grasshoppers for multitude!" And Gideon, who knew not the law that might is roused by communing with the Mighty, answered from his Joseph or human sense of himself, "Oh, my Lord! Wherewith shall I save Israel? Behold, my family is poor in Manasseh, and I am the least in my father's house." But suddenly Gideon rose up with a new sense of himself, kindled by communing with One to whom great works are easy, and masterful deeds simple; and he stood boldly forth before the haughty army of Israel and said, "Look on me!" And all the powerful soldiers of Israel acknowledged Gideon's right to untried martial leadership. He had touched the strength chord that stretches between universal Omnipotence and man's omnipotence. It is the joy chord also;

and the love chord. It is the faith shout of "I Am Victory"
in the hour of defeat.

The mystery of the victorious vision of Esther, and
of David, and of Gideon, is our mystery of victorious
vision by much communion with the same Victorious
Lord.

Scientifically speaking they all cognized from the God
centre of their being, struck back to God centre by much
association with the Unseen Worker, ever offering himself as Almighty Ally without partiality, to all, high or
low, rich or poor, wise or foolish.

As the actinic ray in the sun is the secret of its life-
giving charm, so the joy of the Lord is his strength. And
man may invigorate in that strength till he is royal
magnet to the life of his neighbor, to the health of his
neighbor, to the joyous smile of his neighbor, like Jesus
to the ruler's child, Peter to Tabitha, Paul to Eutychus.

Mystic joy of heart that draws forth the Judah Self
of the sick, is the magnetic mystery which no man on
earth exhibits in the plenitude of its attractive energy,
because no man puts forth from the magnetic "I" centre
of his being, the "I" Jesus hand that came not to be
ministered unto, but to minister.

At the magnetic Spark Centre of our being we are
"the Sun of righteousness . . . with healing in his wings."

As the sun draws the oak tree from out the acorn's
stiff cortex, up through the hard mold, past the tough
grass roots, to centuries of forest life, so the hand that
taketh the yoke from off the Judah jaws of mankind
must move out from a central magnet stronger than
death, stronger than Satan, or the non-alliance of the
sick themselves. (There is no Satan except such non-
alliance.)

No accusation abides at the magnetic centre of our being. If it is love, it is love that ministers, asking nothing in return. It can wake to such might by our alliance with the Mighty, that it draws forth the integrity that hides like a giant asleep in the hearts of all men. The recognition of giant integrity waiting to be awakened annuls the disposition to condemn. There is no time to parley with would-be hinderings. The main business engrosses: *"It is your will—It is my will—It is God's will—*Come forth!"* And even Lazarus may not resist that King's command.

It is lode-star to the Logos of man. It draws forth the axiom of his Sonship to the Father. It wakes the key word to his diviner destiny:

"There is a message speaking within me from the heavens; Whispering within me even in my sleep."

It stirs the victorious *I am That!* "I am a writer," said Balzac. Was not this word the chord to which his life vibrated? Did not that word act like a chain pump to the pool of genius hidden in his bosom? But there is a deeper and farther reaching adequacy lying at the root of our being than Balzac spoke.

There is an axiom that is bone of our bone and sinew of our sinew. It came with us when we hailed hither from our heavenly Father's bosom. If we speak it forth, word for word and letter for letter, we are the sunshine and gladness of every life with which we associate. No situation daunts us. It is the laughing topaz stone of fearlessness, because nothing hath power to hurt its self-existent *verity.*

No wonder that the true topaz stone, emblem of unkillable joyousness, is almost priceless in the gem markets of the world!

Inward buoyancy, kindling into speech, dissolves danger, animosity, failure. It puts them at disadvantage: "I have set the Lord always before me . . . Therefore my heart is glad—my flesh also shall rest in hope," said David. He got his gladness of heart from setting the Lord always before his face. And the confidence of his flesh rose to such vigor that he set Goliath to one side, though a whole nation trembled at sight of him.

To this day the words, "I come to thee in the name of the Lord of hosts," which David used when he faced Goliath, have something of the victorious self-executive David energized into them, even when the genesis of their energy is not noted. They are living examples of people who, though they have not set the Lord always before their faces, and therefore have not stirred the joy spark of their being, yet have been made mysteriously triumphant by repeating David's words, "I come unto thee in the name of the Lord of hosts."

The words turned their vision to the hard problems of their life as being met by that Unseen Ally who saith, "I will contend with him that contendeth with thee." It is a continuation of the shout of the Israelites following on after the Ark of the Lord, as it led forward into the ranks of their enemies: "Rise up, Lord! and let thine enemies be scattered; and let them that hate thee flee before thee!" Their vision toward an Unseen Ally moving with irresistible might before them, worked havoc among their enemies; took the yoke of enmity off the jaws of their rightful safety, drew glad security to their plain view. "I was to them as they that take off the yoke on their jaws."

Haggai promised that the "Desire of all nations shall come." He meant that the glorious Lord ever facing shall be acknowledged, recognized, sound, free, wherever we look. "Like attracts like." Magnetic gladness of heart attracts gladness of heart. This medicine divine stays waiting high watch, self recognition, joyous association with the Lord I Am of our neighbor, fearless of his hurting power, fearless of all hurting powers, heavenly hosts for antiphons—

> "Harvest from seed that was scattered
> On the borders of blue Galilee."

"He that overcometh will I give power over the nations." To overcome is to come over. To come over or above our circumstances and bodily feelings, by looking above to Him ever beholding hitherward, is to identify with power above the nations and combinations of delusions, as divine magnets to the strong, glad, free Self, even of multitudes: "Great multitudes came unto him, and he healed all."

> "From many an ancient river,
> From many a palmy plain
> They call us to deliver
> Their land from error's chain."

"Him that overcometh will I make a pillar." *Pillar* is strength. "The joy of the Lord is your strength." Joy gives mysterious potency to the hands visible, and to the hands invisible. To stretch forth invisible hands and touch responsive strong Spirit, and say, "Come forth!" is to go into all the world and "lay hands on," that multitudes may come forward repossessing their own wholeness. "And they heard a great voice from heaven, saying unto them, Come up hither. And they ascended."

"And Joshua . . . was full of the spirit of wisdom, for Moses had laid his hands upon him." "And when Paul had laid his hands upon them, the Holy Ghost came on them, and they spake with tongues, and prophesied."

Though the Angel of man's presence has shown ages long renitency to wails of woe, it springs in full-orbed splendor forth to greet the sceptred word of the joyous Shiloh spark of the New Healer—*Qui mittendus est*.

With vision cast downward, Mary was accusing some gardener of hiding her Lord; with upward vision she beheld the Lord, and no guilty gardener. She brought forth the Angel out of the heavenly spaces, and forgot that the guilty gardener ever existed. They shall "forget . . . misery . . . as waters that pass away."

Thus with upward vision does the delusional age end, and the Reality age appear. Upward vision draws up from our native root of sincerity. By upward vision oft practiced every human being starts anew from his Sincerity Root, and drops the dry branches of lower attention, which have formulated into nervous prostration, decrepitude, poverty, discouragement, apprehension, and flourishes forth from his Sincerity Root of Vigor Everlasting; his Bottomless Life-Pit; his Healing Joy-Fountain; his Winning Bosom-Spark; gathering the Angelhood of the universe to companionship. Is it not written, "Ruling all nations with rod of iron," or with the irresistible pulling strength of magnetic iron? "I, if I be lifted up . . . will draw all men unto me."

So, with divine allurements, again at the alphabet of Mysticism, we are brought with David to "lift up our heart with our hands unto God in the heavens," with sweet commandment, praying, "Turn thou us unto thee, O Lord . . . renew our days as of old." This is the heavenly exhortation that makes our message as ointment poured forth. This causes the responsive nations to say with David's son, for the Great Ministry's sake, "Draw me, we will run after thee—Many waters cannot quench our love" of thy wondrous Word!—that Memra which created the world; which brought Israel out of Egypt; which brought the miracles recorded in the deep book of Exodus!

"From the sixth hour there was darkness over all the land unto the *ninth* hour." Is it not always understood that darkness is ignorance? The sixth hour sets forth the law of spontaneous, or volitional calling after the Unknown and Invisible, to let It take possession of us after Its own fashion. The seventh and eighth tell of forging ahead for our neighbor's Unknown and universal Best Good to take possession of him, manifesting after Its own fashion. But the *ninth* in all ages treats of the knowing, the cognizance, the independent awareness by our own apperception, of the reality of the High Best Self of our neighbor, the great Angel of His Presence:

> "What truth when number *nine* we see
>   Should we remember most?
> The orders it should call to mind
>   Of all the heavenly host,"

the tantibility of the immortal and unspoilable Self, clothed, housed, fed, strong, glad, free, as the outward sign of our recognition of the oneness of his Judah and Joseph self. "Inasmuch as ye have done it unto one of the least of these, ye have done it unto me." This closes the age of darkness, of the hiddenness of the Self, and gives tactile value to the Angel of God's Presence speaking to us and dealing with us, whenever a human being speaks to us or faces us. We appreciate that as there is but One Supreme Self in the universe, ever inviting each self to transcend itself by recognition of its own One Supreme Self, we are all brethren, because our neighbor's One Supreme Self is our Supreme Self also. This is the genesis of the injunction to love the neighbor as the Self. It is the "Brahma Self, One Self, thou that Self," of most ancient teaching.

Cornelius, a Centurion in the Roman Army, had lifted up his fearing vision Godward till the Angel of the High Supreme was tangibly present before him, conversing with him as man to man, lovely, masterful, wise. "At the *Ninth* hour of the day" it was, as we read, Acts tenth chapter.

"At the *Ninth* hour Jesus gave up the ghost." He dropped the sense of a body or mind differentiated from the Supreme Self. He sent forth his divine influence untrammelled by form, to be contagioned by all men world without end.

The Levitical law of Hebraism reads that man shall eat of the old fruits till the *Ninth*. That is, we must stand by what we are taught as gospel truth of appearances, till we know to speak by our untaught Self. Peter is trying in his Second Epistle, to tell us that we have a native word of prophecy, "whereunto (we) do well that (we) take heed . . . until the day dawn and the day star arise in (our) hearts." Day star is lodestar, magnetic centre, where the Original Knower in us waits our spoken *I am That!* and our neighbor waits our spoken *Thou art That!* and the very world waits our *"I decree!"*

"Look upon Zion," says Isaiah, "there the glorious Lord will be unto (thee) a place of broad rivers and streams . . . And the inhabitants shall not say, I am sick: the people that dwell therein shall be forgiven." Surely this is a precious treasure, this topaz stone of independent recognition, and independent speech. What reactions of new wisdoms lie quiescent in us till our eyes do look on Zion, and the Lord Supreme be the place of our vision's glad rest! What consort with Angels our neighbors shall experience by our right view of them, as Mozart experienced new music by viewing angelic choirs!

Every other way of viewing our fellow-men exposes us to his shadow system's discordant traits. Jeremiah speaks of a whole nation as having eye failure because of "vain watching toward a people that could not save."

Balaam of Mesopotamia caught a cursing mind from associating with Balak king of the Moabites, who was filled with the cursing mind. So Balaam was willing to curse the Israelites or any other people Balak might suggest. King Balak had a psychological influence over his associates, like many powerful people of our own time. And Midian, his envoy, who also was under his spell to defeat whomsoever might oppose him, right or wrong, added his vocal persuasions to the king's forceful psychology, and when Balaam got well into their swing he prayed God to make him strong to sweep the Israelites off the earth. After a night's communion with the I AM on high, he opened his mouth with sublime words of blessing and fore-telling for the Israelites:

"There shall come a Star out of Jacob, and a Sceptre shall arise out of Israel."

Balak and the princes of Moab, with Midian, angrily reminded him that he had been hired and influenced to curse and not to bless; but Balaam, still under the bright sway of High Wisdom, caught by a night's communing with The Transcending, could not enter again into their psychological aura, and despite their threats, being as it were in a trance, having his eyes still open, he pronounced the words of the High Deliverer:

"How goodly are thy tents O Jacob, and thy tabernacles O Israel! Blessed is he that blesseth thee—How shall I curse whom God hath not cursed? or how shall I defy whom God hath not defied? For from the tops of the rocks I see him, and from the hills I behold him: Lo, the people shall dwell alone, and shall not be reckoned among the nations." Balaam saw the life of the Jews on earth three thousand years ahead of him, even up to this day, when the Sceptre of the Israelitish stock, the Star of the Son of Jacob is the morning magnet of the whole earth—

"The star that shone over the manger
Now covering the earth with its light."

"He that overcometh, and keepeth the word of my patience, I also will keep him." Come over the personal hardships by looking up. Did not the Great Prophet say: "In the days of world war and universal lamentation, *Look up*" (St. Luke, 21st chapter).

Keep the Great Name, or the Great Word, as set forth in the first study, and prove the promise, I will keep thee from the hour of temptation."

A servant girl feeling the spark burning in her breast said, "I must be a missionary. With God all things are possible, and he has called me to be a missionary." When the mission board received her application they told her she had not the credentials for foreign missionary work. But nothing can down the flame of the *I Am That* when once uttered and kept as precious truth! Is it any surprise to be told that when delegates of different foreign mission stations were in the United States, and heard her sing—

> "Waft, waft ye winds the story,
>   And you ye waters, roll,
>   Till like a sea of glory
>   It spreads from pole to pole"—

and

> "Still, still with thee
>   When purple morning breaketh,
>   When the bird waketh
>   And the shadows flee,
>   Fairer than the morning,
>   Lovlier than daylight,
>   Dawns the sweet consciousness,
>   I am with thee,"

they invited her to go to foreign shores with them, telling her plainly she could win more hearts with her singing than they could win with their preaching?

Keep therefore the high watch and the wonder-working Name, till the spark *I Am That* lets loose its executive splendor with its "I will and I decree."

> "Such are the Elect,
>   Who seem not to compete or strive,
>   Yet with the foremost still arrive,
>   Prevailing still;
>   Spirits with whom the stars connive
>   To do their will."

Speak from the Spark Dominant and see its dominion! "For out of Zion shall go forth the law, and the word of the Lord from Jerusalem," as Isaiah prophesied.

How reviving to mankind is the voice of the God-Will waking the silence of ages with the sceptred word addressed to the True Self of our neighbor: *It is your will to be glad, and strong, and free! It is my will that you be strong, and glad, and free! It is the will of High God that you be strong, and glad, and free!* Walk with me, acknowledging no other voice than the voice of The Supreme now chording all wills into Its One Will. I Am That will—Thou art That will—God is That will—There is but One Will!

"Man has become weary of his thoughts, and seeks for higher power to free him from his mental prison."—COREY.

"Look unto Me"—"I will show thee great and mighty things which thou knowest not."—JEREMIAH.

Exaltation of the mystic Visional Sense wakes recognition of our own hidden God-Seed; and inspiration fans our God-Seed to full exhibition as children of the Promised Golden Age.

E. C. H.

X

MINISTRY

Ten is the number of the Light. It is the Sephiroth giving birth to everything. Ten is the number of the memra, the secret word of the Mystic Self, the hidden Knower in us all waiting bold definite speech to coincide till the without shall be as the within. Ten is the I, Jehovah Self-Providing. I, Jehovah Jireh, Jehovah-nissi, Jehovah-tsid-kenu. "I, the sanctuary in the midst of thee in all places whithersoever thou goest." I, the Secret Word, the Secret Warrior, the Secret Music. I, that Being declared and *none else,* "show thee hidden riches of secret places."

Aristotle B. C. 384 found that there are but ten ideas in the world as there are but ten numbers. All the religions and philosophies and reasonings of the world but ring the changes on ten ideas, as all the mathematical calculations of the world ring the changes on ten numbers.

"With ten words was the world built," reads the Zend-Avesta. The ancient Hebrews declared that *Yod* the tenth letter of the Hebrew alphabet is key to the divine language some time to come forth from the "I," hidden man of the heart, angels, authorities and powers then to exhibit themselves as forever subject unto that divine language.

Only one *ten* character has ever yet appeared on this earth. Notice Him saying, "All power is given unto Me." "I have overcome the world." Ten is the Rose of the World coming into bloom. Ten is the Wonderful Mother, Desire of All Nations, secret of Protecting Love. Ten is Symbol of Resurrection. "And the people came up out of the Jordan on the tenth day." "Then shalt thou cause the trumpet of the jubilee to sound on the tenth day of the seventh month."

Make note of the key speech of the hidden Light: I am the Lord, I change not. I give power; I take away power. I gave power to death; I take away the power of death. I gave power to defeat, danger, deafness, disease; I take away the power of defeat, danger, deafness, disease. "All power is given unto me in heaven and in earth." Here I'll raise my Ebenezer, my Rock Eternal. I, the Lord that change not.

"I consulted with myself," said Nehemiah. "And I rebuked the nobles and the rulers." I had given power to nobles and rulers. I took that power back to myself, and the nobles and the rulers obeyed Me. *"I have overcome the world."*

It is plain to be seen that all who are giving power to death, deficiency, debt, have not spoken from their memra, their Secret Lord Word rebuking such nobles and rulers. Notice that it is *"I"* being declared and *none else* that take away the power of those rulers on this earth, danger, defeat, death.

Everybody seems to be declaring for the powers given to these rulers, that they may keep on with their powers, not denuded of their powers; but the promise is unto all mankind, "My spirit shall not forever be humbled in thee." "I will rise, I will be exalted."

This is the day of the *Ten.* It is the number of the Great Resurrection. "And the people came up out of the Jordan on the tenth day."

Number *nine* stands for joy in our own responsive God point, all peace and plenty.

The Jews of old had the joy of peace and plenty. They have it now. They offered it to the early Christians as the way of their God with all people according to the inspirations recorded in their Sacred Books. The inspired Jesus of the Jewish religion offered it to the Christians, saying, "Search the Scriptures." But they ran away from inspirations of the Sacred Books and got up their own text, "Poverty and Penance." So the Christians have had to struggle with poverty, working themselves day and night, and practicing some way of penance for success in fighting poverty whenever success has crowned their strenuousness; or cowering under poverty or penance all the days of ignoring Peace and Plenty. They have rejected the generous beneficence of "Him who giveth riches and addeth no sorrow," whose "yoke is easy, and whose burden is light."

Number Ten calling again our attention to our native majesty brings back to the joyous God-Self Its original "I, The Lord give and I, the Lord take away." This tenth lesson teaches us to start over again at our hidden memra and sound it forth till the earth shall hear and repeat the anthem, singing with the four and twenty elders of Revelations: "We give thee thanks because thou hast taken to thyself thy great power, and hast reigned."

The silent memra or vach language of wordless knowing deep within us all is that language coming to speech soon after being noticed as incontrovertible verity. "Look to the rock whence ye are hewn." Look to the hidden centre where the Inmost Lord and the Highest Lord is one Lord.

Victor Hugo found himself being taught by his hidden memra or original genius. If we go back far enough into our own crypt or hidden Super Self, we find our own genius like Victor Hugo's waiting our bold expression. Victor Hugo wrote about the mysteries because his hidden Self knew the mysteries as our hidden Self knows mysteries—Mysteries which we may tell forth as new music or new life comes to register on our outer forms; as everything we declare soon registers on our outer forms.

How can it as yet be told so that the world will practice it, that we can willfully lift up ourself free from ourself, till lightness of weight and flawlessness of bodily condition are registered on the *locus standi* or bodily self we manifest to the world?

How can it be told so that people will practice inbreathing those white breaths ever near us which contain the quickening of our vital interiors till

"Come Holy Spirit, heavenly dove
With all thy quickening powers"

is to us no longer just a rhythmic figure of speech, but a white tonic to be drawn in by the nostrils?

Who is there to make it a practical reality to us that we have given to disease, defeat, death, all the power they have, and that any time we choose we can take back to ourself that power, till disease, defeat, death cannot be found?

When the cruelty that claims to be inherent in the breast of man gets aroused, it runs away with him. Even when woman, the compassionate of human kind, lets slumbering cruelty get going, she laughs at the anguish of other human beings. So other slumbering traits within us, being unleashed, are sure to get the better of us and run us their way. We give them power to ruin our happiness or augment it. Notice the power we give to people to hurt our feelings. Now we are taught to take back the power we have given them to hurt our feelings, and to start over again, putting something to work for us as Nehemiah took away the power he had heretofore given the haughty nobles and rulers of Jerusalem and started them to building for him the walls of the House of the Lord. When our secret eternal God Genius, the Nehemiah trait whom Jehovah comforts, starts over again with the nobles and the rulers now denuded of their over-flowing power to hurt, we set them into their rightful business of forwarding our hidden lordship's great work of transfiguring the world.

"Be steadfast, unmoveable, always abounding in the work of the Lord," "He that standeth steadfast in his heart doeth well." Stand to the Nehemiah lordship inherent in your breast till it gets going and denudes all cruelty of its seemingly well rooted nativity. "Every plant that my Father, (my original), hath not planted shall be rooted up."

Now we know to take back the power we have given to our emotions to capsize our peace of mind with grief or fear or joy, and to start them over again with new working efficiency as the Tishbite started three mighty kings in the Edomite valley centuries before our era. Now we know that the Tishbite thus ruling his special nobles and his rulers was fore-type of our own giving unto and taking back Emotions' power to hurt, and giving power only to the proper work of setting the miracles of God to showing forth wherever we walk. "I have overcome the world" is our God-rooted plant, discovered upon taking back the power we once gave to hurt and emotion. It proclaims our secret right to rule, no matter how hidden by grief or anger, those terrible emotions winding always into disease and death wherever they are allowed to rule.

Our Hidden Gnosis tells us that to sing as if we were the singer, while knowing that it is the Singer within who really sings, is to have the New Song put into our mouth according to ancient prophecy: "I will sing a new song unto thee." Our Hidden Gnosis tells us that to fight on in life's battles as though we were the fighter, while knowing that it is the hidden warrior who fights, is to have the New Warrior rise up with mystic overpowering so that nation shall never rise up against nation again forevermore. "A warrior in white shall appear in the tenth Avatar" is the promise. The mystic warfare of the warrior in white is already begun. "And they shall not hurt nor destroy in all my holy mountain"—in the Tenth Avatar. "I will work a work whereat all men shall marvel." This is the announcement of the Lord Integrity within me; my Gnosis.

There is to be a new doctrine revealed to man. It is to come shining forth like a new star in an early morning sky; like a view of the now unseen world that lies adjacent, exposed by some as yet undiscovered telescopic crystal.

There is a Throne One above the sheep and the goats; above the pairs of opposites, love and hate, peace and strife, truth and error. That Throne One hath promised, "I will show thee great and mighty things which thou knowest not." "I have given thee an eye divine" with which to look toward Me and then behold My works with thee.

All the doctrines that mankind have fought for and lived by, have been swirled around their views of good and evil, truth and error, spirit and matter—the differentiations of these as desirabilities and undesirabilities, blessings and menaces. With slight attention paid to the inspirations of the upward watchers, these views of the pairs of opposites have flung all people into the opposing currents of joy one moment and anguish the next moment; peace of body and mind one hour, and pain of body and mind the next hour: the choppy sea of human mutation.

The promises vouchsafed to the upward watchers have been enumerated; but upward watching toward the High Deliverer has slipped the practice of man. Therefore the most glorious doctrine is not yet expressed. We must do the mystical Will of the High Eternal, "Look unto Me," an allotted term, to be sensitive to New Ineffable Mysteries.

And ye shall live, and know, and be strong, and nothing shall by any means hurt you, comes to pass by upward visioning. We must give the mystical Will its heaven-allotted term, to experience immuneness from decay, weakness, ignorance, hurting power. There is to be a new doctrine concerning Reality and Unreality: a doctrine that will hit the confidence of all mankind with the self-evidence of the homely axiom, "the whole is equal to the sum of all its parts." Is the doctrine of the unreality of matter as self-evident as "the whole is equal to the sum of all its parts?"

The Hindus have for centuries maintained the *non est* of all external things: "What seems external exists not at all," we read in their Lanka Vatara. "By such doctrine I am not in a pulpit preaching; there is no church, and no pulpit. Jesus never walked in Galilee; there never was any Galilee to walk in; this is a blasphemous belief!" said a great divine before a London congregation.

Coomra Sami of Thibet could make the trees and rocks and hills disappear from before the eyes of Hensoldt the level-headed German traveller. This was evidence enough to Coomra Sami that the trees and rocks and hills were the insubstantial pageant of sense delusion. But when his hypnotic spell was off the German traveller's eyesight, the trees and rocks resumed their wonted appearance, and the centuries-old hills are still standing. The endurance and substantiality of matter were more surely demonstrated by the operations of the sage of Thibet, than its non-existence. And the mystery of hypnosis has never been fathomed. Only the promised New Doctrine can open the sealed books of many such mysteries of human encounter. "The mystery of God should be finished, as he hath declared to his servants the prophets."

Thus far we have arrived, namely, that with the baptism of the alkahest that falls from the Heights, matter and evil undo their laws of action. But it is only by exalting the visional sense that we feel the alkahest that looses matter's grip on the conditions of life and mind. It is therefore while "facing Thee," that our lips open to say with didactic firmness, "What seems external exists not at all."

By the subtle touch of grace descending, diseases unformulate. We are certain that "My grace is sufficient for thee." We see its sufficiency before the words have been uttered forth. The genesis of all ideas is inward viewing, or percept. The greatest ideas are generated by highest viewing. This is not a new teaching. In scholastic metaphysics we read that "the existence of ideas is subsequent to that of perception—and even implies perceptual cognition."

So the doctrine of the formulative power of vision is not the New Doctrine foretold by the prophets, seers, and philosophers, but it closely precedes it, as the stiff insistence of the tremendous power of thought closely preceded the undownable insistence that inward vision gives mystical vitality to thought, and without inward vision the thought is sounding brass and tinkling cymbal.

The New Doctrine just at our gates, to follow close on the footsteps of the living veridity of attention upward, is not one that is now already declared. It is the "great and mighty thing which thou knowest not," as promised by the angel to Jeremiah six hundred years before Christ, and by the angel of the Apocalypse one hundred years after Christ.

It is the hurrying of this New Doctrine hitherward that is sweeping such multitudes of downward watchers out of sight. Did not Jesus promise that in the day of nation against nation, only those who should be found upward watching should be manifestly blessed? And in ancient mythology do we not read that in the coming conflagration only *he who sees* is saved?

No insistence that *I am strong and every whit whole,* can be found exhibiting strength and wholeness if the vision is still upon the body's misery-claims. Only vision above the body can bring back over the track or Tao of the vision the healing alkahests of the Heights. No insistence that *I am victorious* can bring victory, if the inward vision is still resting upon the misfortunes and evil liabilities of existing affairs. Only attention to the Highest Best can work the best into our life lot. "Man alone of all the animals knows enough to seek his high-est good at the highest Source, and he looks to his Maker," wrote the mystically wise Lactantius, tutor to the son of Constantine.

The *ninth* chapter in Mysticism brings to speech the *I am God* that glows with unquenchable might at the centre of every living being. It shows that the Highest Lord and the Inmost Lord is one Lord. It shows that visioning high we vision deep toward Christ in us the hope of glory, whose is the power and the might forever.

This *tenth* chapter in Mysticism drives to the sturdy Job declaration, "I will maintain my *cause.*" So much has been written by the sages of the ages concerning this matter of maintaining a once expressed conviction, that we had better follow up to know the value of standing firm in the midst of the proud waters that sweep over the life of mankind, tempting to non-expression and forgetfulness.

"He that standeth steadfast in his heart doeth well," wrote Paul. As though some excellent work should out-show firm inward ackowledgment. "Beware lest ye also . . . fall from your own steadfastness," wrote Peter. "Thou shalt be steadfast and shalt not fear," said Zophar the Naamathite to the tormented Job. As though steadfast-ness to an inward believing-spark would quench the smoke of fear.

This is, then, the genesis of the urge, *fear not.* That is, if we are steadfast to stand by the central spark *I am God* that never dies in our bosom, fear will take flight of itself. "Thou shalt not fear" is prophecy, not com-mand.

David calls his inward spark, "mine integrity within

me." Ezra calls it 'His Sanctuary." Ezekiel calls it the
Lord's "little sanctuary." They all agree that it is the
indestructible element, "the rock of mine heart" in all
mankind. And they all notice that it is the unworded
Logos ever patiently waiting for the worded insistence
of each of us, without respect of persons.

Any expression of the integrity rock, the believing
point at the silent depth of our being, acts exoneurally,
or beyond the limits of the bodily presence, to affect the
mind and nervous system of our associates. Did not Dr.
Mayo of King's College, London, feel the exoneural influence, or the influence beyond the limits of the bodily
presence of certain people? Did not Pliny the elder notice
that some people's bodies shed forth medicinal influence?

Inward God-believing stays unspoken through the
lifetime if one does not choose with firm choice to speak
it forth. With expression of the Integrity-Word "the Law
goes forth from Zion, and the word of the Law from
Jerusalem." Nothing resists firm inward maintenance of
the Rock-Centre-Truth that lies so still, native to us
all, waiting and waiting bold, silent utterance. "All the
people bless those who dwell in Jerusalem," wrote a
Hebrew prophet. He knew that those who stand by their
noble convictions exert far reaching influence.

"I believed, therefore have I spoken," said Paul,
whose exoneural influence woke multitudes to new life
in his generation; and whose influence proceeding forth
to this day stirs the faith spark to shake the emotions,
as Felix shook; or stirs a bold, responsive *"I Believe,"* in
others; strong in effect as was the convincing energy of
the man at the Siloam pool.

Confession of faith is an invigorating practice. "I had

fainted unless I had believed," said the king. "I consulted with myself, and I built the walls of Jerusalem," reported the cup bearer of King Artaxerxes Longimanus, when surrounding enemies crowded him back to his bedrock executiveness.

The Throne place in man is his *Rock I Believe*. It is strange that men have not noticed their own unbreakable original *I Believe, I the Lord,* with its wonderful exoneural masterfulness! "Signs follow them that believe," said Jesus.

The faces of the golden Cherubims of the temple are always turned inward. Cherubims are emblems of thoughts and senses. The senses get their cues and clues from inward believing, either as clouds and darkness of false believings, or shining sparks of true believing. Is it not written that "his own counsel shall cast him down"? Do we not notice in life that the Napoleonic "I believe that God is on the side of the heaviest artillery," finally casts its Napoleons on the sands of defeat? Such a believing is a smoky believing, or a non declaration of Original Believing.

When Nehemiah consulted with himself, he held his dialogue with the Throne Place of himself, not with the clouds and darkness around about that Throne. He struck back to his *Gnosis Rock,* where Irresistible Rulership abides forever in all men alike. Thought and sensation are the cherubims of our sanctuary, our temple. They train together. They experience all things of existence according to the depth toward which the vision turns, from which depth they draw their conclusions. With faces turned inward to the Rock from which we are hewn, they draw from the Gnosis throne, independent of outward encounters; or if self recognition is not to the Rock of "I, the Lord," they draw from the trembling cloud of unknowing round about the throne of hidden Knowing.

At the Throne Place, the Zion Centre, our Knowing and Believing are One. "I know whom I have believed," said Paul, "he is able." This was the believing that Jesus meant; the masterful executiveness of "my integrity that is within me."

Wonderful things are spoken of the Indwelling Gnosis: "Thou fillest with hid treasure." "Let search be made in the king's treasure house." "Who hath put wisdom in the inward parts?"

With high instruction as to the training of our senses and our thoughts toward the treasure house of our kingship Isaiah enjoined, "Look unto the rock whence ye are hewn, and to the hole of the pit whence ye are digged."

Thus without sequential presentation of that which builds thoughts and sensations, all the suddenly inspired of time have set their hand and seal to the precedence of the looking faculty in all accomplishings and all experiencings. And they show how the exaltation of the mystic sense to the heights is ever followed by its sure return to the Sonship Centre, the Unlimited Executive throned in man himself. "Did not he that made that which is without make that which is within also?"

Nehemiah reported that only one out of ten dwells at his own centre, even among those who have been told the Great Gospel. And he seems to know why all the people bless those who willingly offer themselves to look toward their own Jerusalem centre. It is because they only of all the children of the gospel send exoneurally forth their integrity's resistless pulling power on waiting health, life, joy, wisdom. *"Canst thou draw out leviathan with a hook?"*

It was Coomra Sami's firm believing that shed blanking over the German traveller's senses. He believed in his own power to give manifestation and his own power to take away manifestation. But his firm believing was not the original Gnosis or holder of his own power at his throne place, for the effects of his believing was transient, while rivers of the Everlasting flow from the hid place of our original knowing, our giving and taking Rock.

Behold, I will lay thy believings with fair radioactivity, is what Isaiah the prophet means by promising that our stones shall be laid with fair colors. When he writes of the sure foundation, the tried stone, he means the original Believer and Knower at our centre, with its tremendous exoneural, far proceeding energies, world without end.

To look to the Rock whence we are hewn is to begin to say, *In mine integrity within me I Believe, I Know that I and the Father are One.* This was the "liver-forth" point of the magi of old days; the lever-force, the believer-executive never destroyed in any creature. Whoever has looked to this his own foundation stone, and spoken according to its speech, has proved its working force.

"I, Alfred Tennyson," said the young writer, when all the reviewers called his work "drivel and nonsense, and most dismal drivel at that." And he shut himself alone with his *I, Alfred Tennyson* for ten years. At the end of that time *I, Alfred Tennyson* was pronounced the prince of song from one part of the world to the other; in palace and hut, among the literati and the ignorant.

For "he that standeth steadfast in his heart . . . doeth well."

"I, I, I, Itself I, the whence and the whither, the what and the why, I, I, I, Itself I!" wrote forth one who had looked to the Original Rock—the pit of the unlimited *Self*. And such victories over the world's oppositions as this "I, I, I, Itself I," wrought forth, are the wonder of the New Age.

Mozart felt the music of the land adjacent, and with joyous confidence he offered his inspired pages to a Publisher. "I will not pay you a penny for such music," said the publisher. "Then, my good sir, I must resign myself to die of starvation," answered young Mozart. And he did. Notice what power he gave the refusal of money! Notice what power he gave lack, deprivation! It has been pronounced a lasting disgrace to the land of Mozart's birth, that his death and burial were like a pauper's closing down. But were they? On the principle of the propulsive strength of inward believing, how could his contemporaries hold up against such positive steel needles of propulsion as his dangerous inward dialogue gendered? He did not remember that "for we are made partakers with (victorious) Christ, if we hold the beginning of our confidence steadfast unto the end."

Plato spoke of a handwriting on the liver in man. That was as near as he could express himself on what Job discovered as wisdom in the inward parts. By this we see that Job was nearer gathering his cherubims or senses and mind to the inward Knower than Plato.

Ezekiel the prophet tells how the King of Babylon consulted the livers of his responsive animals to read how it was proceeding with his own outward affairs. The death of Alexander the Great, and also of his general

Hephastion, was foretold by the livers of their animals, acted upon exoneurally by these great men's inward dialoguings. How much thay needed to know that they could take back the power they gave to the liver records and decree life renewing which the next animals would register!

Everything and everybody not braced to independence by looking to his own original Knowing and Believing Rock gets under the projective force of the inwardly stronger among the cloud-and-darkness believers; that is, under the exoneural power of his propulsive neighbors. Much of the healing done in this age is accomplished by the strongest holder of a healing believing. When the man who believed himself aided by the great red dragon of Scriptures cured so many ailing people of England, was his believing based on the original Rock of Truth, or on a stiffly maintained conviction outside his Original Gnosis? Was it not from the cloud and darkness round about his deep *I Know?*

When a certain woman cured so many sick people by the influence of an invisible Indian attendant, was she strong in Original Truth, or sturdily believing outside its Inward Throne? That is, outside her "I gave power to the Indian; I can take back the power to myself and need nothing but my own far extending right decree to cure my neighbors."

"That was the Light that lighteth every man that cometh into the world." Nobody doubts its word when it expresses itself. From deep to deep the world around each answers, *I Know That!* This *Tenth* chapter urges the finding of the *I Know.*

George Eliot declared that she was possessed by an inward demon of despair when she wrote *Romola.* This was her dangerous consultation with cloud and darkness believing, held firmly till its exoneural affected all the people who read the book. They caught a touch of her inward despair. Napoleon's false believing worked with strong influence on his responsive soldiers, till at Waterloo, in Belgium, 1815, his believing that God was "on the side of the heaviest artillery" broke down. For, "Every plant that my Father hath not planted shall be rooted up." Napoleon's plant was never planted of our Father. Our Father's Plant reads, "Not by might (and not by army), but by my Spirit, saith the Lord."

The Holy Spirit is the radio-in-extenso of God. The Holy Spirit revives, instructs, empowers, defends. It is the radio-in-extenso of all who look to their own Inward Integrity Root and rest on its victory-shedding *I Believe! I Know!*

And believing is stirred by seeing. Dr. Evans the great healer, could with inner vision see so plainly the action of the *vis medicatrix naturae* in sick people, that he could say with bold confidence, "You are being healed." And they would hurry into health. He began to see in himself the action toward him of the exoneural energy of some who conceived of him as a menace to their own precedence. Having no vision toward the *Unhurtable Me* at his inward foundation, he perished. The far reaching influence of the mistaken conceptions of his persistent enemies found workable soil in Evans. See how great a defense from damage it is to have a boldly declared inward conviction of the undestroyable *Me*: "He that believeth on *Me* hath everlasting life."

Let us take today to "look unto the Rock whence we are hewn, and to the hole of the pit whence we are digged." Let it be nothing to us how badly or how goodly human activities are secretly or openly dealing with us. The more executiveness they show, the stronger the call for our looking to the Rock where *I Believe in Myself as Undefeatable Lord of the harvest.*

When Paul discovered that Timothy at Ephesus was always breaking down under the secret criticisms of the Ephesian Christians, he wrote to him urging him to take heed unto himself, and to continue in his doctrine; "for in doing this thou shalt both save thyself, and them that hear thee." And obedient Timothy looked to the Rock whence he was hewn, and strengthened up by re-stating his original Believing and his original Knowing, to such a degree that he ceased from ofttime falling into sickness and tears of discouragement, and drew the Ephesian Church with him into an invigorating Centre of Divine Ministry.

Paul wrote to little Titus, the terrified preacher to the Cretans, those celebrated liars of Homer's time and of Paul's time: "Speak thou the things which become sound doctrine—sound speech that cannot be condemned; that he that is of the contrary part may be ashamed—that they which have believed in God might be careful to maintain good works."

And Titus maintained his Rock Integrity so robustly that he stopped the lying of the Cretan Christians; made his own instructions their guide, and his name their watchword for more than a hundred years.

"Like attracts like." "No man can come to me except the Father that is within me draw him." *Father* is sometimes mentioned as Faither—The Integrity Rock of my God Faith, my Believing Centre, my inward-Knowing Point. The artist draws artists around him. The musician draws players upon instruments and singers of sweet tunes. The sound in doctrine draw around them lovers of sound doctrine. The great draw the great. The greatly true draw lovers of truth. The Integrity-healed strike forth Integrity-Health as flint and steel strike forth sparks.

By the practice enjoined by the ninth chapter, we pay attention to the Judah Angel of man's presence; and the Judah Angel of his presence glows and invigorates the Joseph human of his presence, till only the sound, sane, powerful Son of the Everlasting Father stands visibly before us! It is magian ministry to use our power to draw forth the Judah Self of man.

Often the Joseph self, the external mind and bodily frame, show forth the perturbations of readjustment. As the sun breaking through the clouds causes a hurrying and scurrying of the clouds, but the sun wins hitherward its shining face, so the disorders of flesh and mind hurry and scurry for "the Sun of righteousness with healing in his Wings" to glow in face and form of neighbor.

When the sun has won its glorious way the clouds cannot be found. They have been lost in sunshine. Thus all excitements of mind and body disappear. The emotions of fear and anger, grief and distrust, jealousy and greed subside. The pains and sicknesses of the body rage and hurt anew, then disappear.

Whatever of bed rock truth is real to us is soon real to our neighbors. Though their dissensions rage, they are sure to show forth health and strength as signals of agreement. Is it not written that Michael and his angels fought against the dragon and his angels, but the dragon and his angels fought among themselves, till they could not be found? So do the clouds hurry and scurry among themselves till they cannot be found. As nothing of cloud has hurt the sun, so nothing of dissension has hurt the shining *Me* of our neighbor. He feels our recognition of the immortal and victorious Spirit of God glowing through him, and straightway he forgets his old sensations.

To obey the command *Look unto Me,* is to be a stirrer-up of seditions among the sensations and thoughts of our associates, like Paul among the men of Jerusalem, when by carrying the victorious Name in his heart he caused a great uproar among them.

Paul was not found disputing with any man, yet they called him a "pestilent fellow." He did not try to set the people to quarreling, yet he was arrested as a "stirrer-up of seditions." Ezra the reformer, nearly seven hundred years before Paul's time wrote, "I am for peace; but when I speak, they are for war." Ezra could not even speak peace in his secret heart without setting his neighbors to wrangling.

The *secret heart* set for shining peace stirs up the cloud thoughts of men. But peace is a ruler as the sun is a ruler. "My peace I give unto you," said Jesus. And families shall become excited, and friends shall argue, as their former thoughts begin to let go; but I will not cease calling, "Come, ye blessed of my Father!"

Let him that standeth for the might of peace ever abiding as the Masterful at his secret Throne not give away to the turmoils of mind or matter. Let there be firmness in the heart. Keep to the Key Science, as the true singer in a choir of voices keeps the keynote regardless of the voices out of tune. If the true singer loses the key the choir's music becomes dissonance. If the true singer maintains the key all the other voices come to accord. It is inexorable truth, that maintaining the Throne Science is standing at the Success Point of our being.

Standing on our original conviction that God reigns, we find that what seems to be punishment for sins is unformulation of the framework of false notions. Is it not a false notion that Peace is only victorious by killing those who offend against Peace? Yet this is the position taken century after century by the teachers and leaders of the easily beguiled children of earth.

At the siege of Leyden, was the little handful of helpless people obliged to slay the crews of the Spanish battle ships in order to be victorious? No. The people of Leyden prayed till they struck back to the rock place of their own original confidence in Victorious Peace, and the menacing Spanish fleet disappeared.

Is the law of the Rulership of our Throne Place abrogated? No. "If my kingdom were of this world, then would my servants fight" is everlastingly true of the Throned Christ in men. The time is ripe to stand for the *I Christ in you the hope of glory;* the mystical mastery of *Mine integrity within me,* taking back the power we gave to fighting for peace, and giving forth peace for fighting. "If thou wouldst believe, thou shouldest see the glory of God."

In the presence of Death, the supposed world conqueror, there was a greater Master feeling the far radiance of the Living Father, his own Original Self. And Death let go its clutches. And the soldiers of the Roman legions fell on their faces before him. "Watch ye, stand fast in the faith, quit you like men, be strong"—for "the angel of the Lord doeth wondrously."

We all believe in this divine law of action at our own

original Believing Point. *Let us inwardly say so!* And let us maintain the secret word of our own Original Knowing till it demonstrates beyond ourselves by its own exoneural or far reaching energy.

When the head of a nation declares that the only pillar upon which his nation rests is his army, can he expect his pillar to endure? Must he not eventually feel the army-pillar cracking under the weight of a nation's leaning?

The leader of a nation has a great opportunity. The more he is revered the more gladly his people believe as he believes. Let him declare with the Original Believing common to himself and his people, as the King of Judah declared before his army, leading them to incredible victory, "Believe in the Lord your God, so shall ye be established; believe his prophets, so shall ye prosper."

If the whole world is hidebound in its cloud and darkness believing, the whole world will be in hurrying and scurrying of unformulation when the silent few shed forth the mastering beams of their firm confidence in the Lord as the *Only Defense* of the world. They know their kingship because they know that the Highest Lord and the Inmost Lord is one Lord.

"But call to remembrance the former days, in which, after ye were illuminated, ye endured a great fight of affliction," wrote Paul to the Hebrew Christians hidebound in their old believings. Felix trembled at the preaching of Paul, but did not tremble hard enough to unformulate defeat and death. The Swami Vivakananda trembled in the silent presence of a believing Guru, till he unformulated all the false notions that hugged him down. The notions made him sick in body and hysterical in mind as they hurried to disband. But the Sun of the Guru's knowing that *I in you Almighty* won its way, and made a shining light of the youthful Hindu.

"The hand of God hath touched me," said Job, when the perturbations of his mind and body and affairs of life threatened his utter destruction. But he came out as Job the triumphant Lord of Ur.

"And I, Daniel fainted, and was sick certain days." What made Daniel tremble thus into sickness and unconsciousness? An angel had touched him, reaching to the integrity within him. And Daniel rose up greater than the mind and body of his former self, and stood forth as revered head of the astonished magi of Babylon.

Balaam's body pained him and his mind was angry and argumentative. What had stirred the thoughts of Balaam's mind to fury and the state of his body to pain? The presence of a shining angel standing in the way. Balaam calmed down upon agreeing with the angel's message, and lost himself and his pains in tranced sight of the Star of Jacob, the Bright Messiah of time to come.

Many a sickness and many a mental resentment would subside, leaving a Believer with wholesome health, if the pangs of their unformulation had not frightened the preacher of the truth. See how much more radio-in-extenso the stiff believer in his troubles exercises than the gentle Timothy-like preacher of the truth sometimes brings to bear. Let the Timothy type retire him to his closet, and looking to the Rock, whence he is hewn, the original *I Believe* and *I know* in himself, let him state again silently and firmly: "I gave power to opposition. I take back to myself the power I gave to opposition."

"Could we but stand where Moses stood,
  And view the prospect o'er,

Not Jordan's stream, nor death's cold flood
Could fright us from the shore."

The rough River Jordan was the dividing line between the Hebrews as slaves in a foreign land, and their establishment in their own country as a free nation. There is a dividing line between being caught in the machinery of the terrifying non-statement of inward knowing, and standing free born and masterful on the silent Insistence of *What I Inwardly Believe and know!*

This is our Jordan *Tenth.* Notice how "the people came up out of Jordan on the *Tenth.*" Notice how the firm Joshua told the people to sound the trumpet of jubilee on the *Tenth.*

All people with some inward believing from which they never swerve find things of nature and affairs of life entering into league with them. Peter was told that heaven would stand by his bindings and his loosings, making his speech to work miracles, because his original believing had expressed itself. Peter sensed his own *I gave power and I take away power.* He felt what Jesus had announced: "I have overcome the world."

Elijah the prophet astonished the Zarephath woman by the miracles his speech effected. Others had spoken the same words but they had accomplished nothing. Elijah had held the divine dialogue unswervingly: "I, the man of God."

The wife of Manoah was not overthrown by her husband's declaration that they had seen God and therefore they must die. She stood sturdily by the plant of our Father's planting, *that to see Him is to live.* And she brought forth Samson, the strongest man Israel ever produced. He had the sternest legislative ability and the boldest executiveness Israel had ever confronted. The anarchistic majority were held in awed subjection by his masculine masterfulness. And the marauding nearby tribes fled at sight of him.

Something Samson-great is always the outcome of boldly maintaining the *Gnosis* or deeper wisdom born with us. Job accomplished great things by steadfastly beholding toward the face of the Transcending Almighty and daringly proclaiming *Thy hands fashioned me—Is not the root of the matter in me?*

I in you is the plant indestructible. It is the Rock of Refuge in the day of calamity. It waits the Moses smiting of my ofttime stern insistence, till what *I am in mine integrity within me* masters for me wherever I walk.

Our own secret text gives its color, clue and climax to our life affairs. *"I believe God reigns,"* said a young man, with his heart in his voice. Every business he undertook failed under his hands without his once flinching from his rooted conviction. At last one failure more disastrous than the preceding nearly choked off his "I believe!" But he rallied, and uttered it forth again. A miracle suddenly turned the scales. Suddenly he was as victorious as David of Judah. If he had neglected that last utterance of his believing he could not have entered into the miracle, for it was by looking up and protesting, "I still believe God reigns," that he was suddenly given the miracle of new heaven-prospered conditions.

"In the morning sow thy seed," said Solomon, "and in the evening withhold not thine hand; for thou knowest not whether shall prosper, either this or that."

Unswerving maintenance of inwardly planted truth to the time of its outer demonstration, is symbolized by the chrysoprasus stone—the apple green stone of a fresh new earth; or new outward conditions among our associates and their affairs: the invariable resultants of our silent dialogue, "I, the Lord."

No matter what adversities face us, they are breakable under the ministry of Inward Truth invariably stated in trouble or in peace. "If thou faint in the day of adversity, thy strength is small." That is, great strength will not show forth by standing to and practicing flinching.

"By the rivers of Babylon, there we sat down—we hanged our harps upon the willows—for there . . . they that wasted us required of us mirth—(but), If I forget thee, O Jerusalem . . . let my tongue cleave to the roof of my mouth."

Though savage enmity and perpetual misfortunes seem our portion, let us not forget the Jerusalem Rock Self whence we are hewn, with its gushing waters of goodly miracles, our rightful portions forever.

We must not forget our Zion Lawgiver though our emotions seem nearly to wreck us. "Get thee to the prophets of thy father, and to the prophets of thy mother," cried angry Elisha to the King of Israel who had insulted him. But suddenly Elisha remembered that his mission was to bless; and he blessed King Jehoram and his two friends with heavenly victories.

There is a voice within us inaudible till made audible by our definite insistence. And this voice shaketh "not the earth only but also heaven," said Paul, "and signifieth the removing of those things that are shaken . . . that those things which cannot be shaken may remain."

"And they said unto him, Master, where dwellest thou? And he said unto them, Come and see. And they came and saw where he dwelt, and abode with him that day; for it was about the tenth hour," *I, in you.*

From the beginning it has been written that the priests of Integrity shall wear clothing (or aura) of salvation, till all flesh shall transform and all mind shall renew as heavenly exoneural of their definitely stated inward conviction, not of sin but of the Mighty-Believer in each man, Lord of Its own environments, Knower of the Final Doctrine; Seer of Itself as Self-Existent Deity, giving life to the faint by taking away the downward crowding power of faintness and decreeing strength; inaugurating Peace by withdrawing the power of war and decreeing Peace.

These signs shall follow him that consulteth with himself, taking to himself his scattered power to reign as A New Worker in A New Dispensation. "Take heed therefore unto yourselves, and to all the flock, over which the Holy Ghost hath made you overseers."

The Eleventh and Twelfth Lessons constitute a science by themselves; a science for the most part unintelligible save to the awakened practitioners of the directions of the First Ten Lessons.

The Eleventh and Twelfth Lessons are the Eschatology of the Mystical Science.

E. C. H.

XI

MINISTRY

Keep hard to one science till you master it. Light flashes on any subject or object by much attention to it. The disciples of the Christ preached One Name only till it flashed into magian effulgence through them and they founded that mighty league the world has since acknowledged as destined to endure forever. They believed that "He that is faithful over a few things, I will make him ruler over many."

King Solomon stood to the building of one house till in the eleventh year it shed abroad such an influence from the unseen world to which it was pointed that all who looked to it were set at liberty. Jonah in the darkness of his dreadful prison, looking often and often to the temple's distant dome glistening under a noonday sun his outer eyes saw not, left his dark prison, hurled suddenly toward the Heaven-charmed house of Solomon.

David knew that whoever would look straight to the Unseen Highest would be set at liberty from his weight of flesh and his weight of foolishness. "They looked unto Him, and were lightened," he declared. So true has this ever been that people have actually found themselves dropping corpulence by seeking Ain Soph the Beautiful Countenance of the Absolute, above thinking and above being. They have also found themselves lightened of their cap of ignorance, knowing wonderful and happifying laws by ofttime and ofttime answering back the wooing Supernal with wooing responses. "Thou . . . hast made me wiser than mine enemies," cried David, after daringly announcing that his "eyes (were) ever toward the Lord."

Looking toward a sky beyond our visible sky, the ancient Norsemen saw the Adam and Eve of the new world which should once more arise after the twilight and destruction of the gods. They caught Norse names for the Adam and Eve of the world to be—Lif and Lifthrasir.

Looking to the sky above the sky, the Mayans of ancient days found the origin of mind. It was a goddess. They named her by a Mayan name signifying, Mother of the Nature-Mind.

Gazing high, Hesiod the Greek saw rushing spirits scattering gifts among the sons of men, which the sons of men took no notice of except when the gifts rushed into them here and there, as stars sometimes collide with each other. Hesiod did not seem to discover that the sons of men may woo the rushing gods to grant each its own kind of gift to the wooing one. Other Greeks tell us that as the worshippers of Mercury wooed Mercury to grant unto them prosperity, and the worshippers of Morpheus cried unto him for sleep, lifting up their hands with their eyes, beckoning and beckoning Morpheus till in his whiteness he folded them in sleep, so every nineteen years the higher-wooed Apollo came into sight to grant happy harvests and beautiful offspring.

Some ofttime watcher toward the Heights above the skies, to what a great king once called "the glory above the heavens," is to catch a new language. It is not to come to the greatest linguist among us for he has had his gaze on men to catch their intonations and accents;

or on books to teach him grammatics and word derivations. The new lingua shall be the speech of right judgment to some second Abraham facing Unspeakable Judgeship, baring his head to the Inscrutable Unseen, meekly declaring, "Shall not the Judge of all the earth do right!" And a great multitude shall speak his language; a great multitude whom no man can number; for it shall be the speech of heaven and earth as one speech, whose words shall instantly accomplish that whereunto they are sent.

If Apollonius could speak six languages he had never studied, because of his much talking to invisible gods, how wonderful must be the discourse of "the Great, the Mighty God, great in counsel and mighty to work" for that new Abraham who much and often notices that the Judge of all the earth is speaking divinely executive words never before heard on this planet!

The very name *Abraham* means *a great multitude*. What a wide swath shall the new Abram or Abraham cut, he who holds his conversation high!

Notice how broken up and wildly dissatisfied people are if events and circumstances run into unwanted cosmic currents which no human being seems to be able to stem! They are no Abram or Abraham of the New Dispensation. Only the old language of grieving and rebellion advances forth from them toward us. The psychometry of them is old "Ephriam . . . like a silly dove, without heart," as Hosea the prophet explained.

Let us notice the comrade Abram drew toward himself by agreeing that a Wonderful Judge is handling the universe: Melchisedek visited him! Origen said Melchisedek was an angel. Martin Luther and Melanchthon said he was Shem a Survivor of the Deluge. Epiphanius said he was the Son of God appearing in human form. The Jews said he was the Messiah made visible. He brought out bread and wine to Abram, and blessed him!

So he who holds his conversation high shall comrade with noble visitants, and cause their inspiring instructions through his speech to ray forth New Truth throughout the earth. "Is it well with thee?" sayeth the High Judge. "According to Thy judgment it is well with me, therefore it is well," he shall answer. Thus he grants the judgment of the Great Judge to strike fire with his own judgment, as bamboo stick striking bamboo stick sparks fire, sparking him into new life conditions. So, and only so responding, can man swing in with the advancing New Order, rejoicing with New Rejoicings, shedding across the earth morning beams of the New Age. Let us isolate with the Heights, that our ministry be of the Heights, new, far reaching, irresistible, after the order of Melchisedek!

Attention to the Overlooking Vast Highest wakes Self recognition; calls attention to hidden Self; what Novalis the poet friend of Schiller, called our transcendental Ego. Some other long-ago mystic Germans noticed that as they were glancing upward they found the Heights gazing toward them. Then they declared, "God's sight with which He sees me is the same sight with which I see Him."

As all invisible operations clashing together form visibles, so right judgment is made visible in speech and action by looking toward the Judge "that shall be in those days," as Moses foresaw for us. "He shall bring forth . . . thy judgment as the noonday."

We have certain invisibles to glance toward which

cause undesirable speech and action to formulate. Notice John Calvin upward watching toward his own flying imaginations, and seeing them smiling on his actions as he burned the good Servetus for disagreeing with him! Beza an adorer of Calvin's imaginations declared that all who opposed Calvin ought to be hanged.

But "When thy judgments are in the earth, the inhabitants of the world will learn righteousness."

"Thy judgment was unto me a robe and a diadem."

"Thy speech bewrayeth thee," the officers said to Peter, who had been with Jesus, the New Teacher, in Jerusalem. Doubtless when Peter spoke of high watch they knew he was a Christian and could probably raise the dead.

Only attention to the Highest wakes Self recognition that rises to cross the bar above prejudices and false judgments. Across the bar Truth is the only language. Truth according to that language has promised to stop the discords of human minds and its angry sensations. We are all candidates for that victorious truth not yet spoken.

There is a Health zone facing us. Notice that the sick who have not crossed the Health zone have to spring above the bar, laying hold of Health, or they must stay under, this side the Health zone. David lifted up his hands with his eyes and caught on to the Hands of Health stretching toward him. "I cried unto thee and thou hast healed me," he acknowledged. So also Peter laying hold of the Hands beyond the bar stretching toward him, crossed above into Health, taking many people along with him.

There is originality beyond the bar. Let us lay hold of Original Knowledge stretching hitherward and be drawn up into new quickening doctrines. Let us speak with New Tongues. The quickening wisdom hither streaming shall cause those people who like Sisera are planning to do the world mischief, to suddenly forget their plans, and themselves be drawn high across the bar into Wise Peace.

Let us acknowledge High Judgment. Let us cross the bar into that state where we are not afraid. Judging this side the judgment bar we are often afraid; but "Whoso hearkeneth unto me shall dwell safely, and shall be quiet from fear of evil."

Come, O people! Let us all together cross the upper bar, laying hard hold of the Hands that stretching toward us draw us into fearless wholeness!

"We went through fire and water, but Thou broughtest us out into a wealthy place."

"He sent from above; he took me; he drew me out of many waters."

Let us all together be born into the Above! The Apostles called high in their day, led off by Peter the ardent: "By stretching forth thine hand to heal," he cried, "signs and wonders may be done by the name of thy holy child Jesus." "In strength of hand the Lord spake thus unto me," reported the son of Amoz, brother of the king: "The people that walked in darkness have seen a great light."

The executive faculty in man is his inner visional sense. Wherever this visional sense looks other senses are pointed and insist likewise with it. Looking upward to a sun it cannot see, the oak tree rises into a higher country all sunshine, wind, and rain to nurture its awakening greatness of body, limb, and leaf. Enduring as seeing the Sun of righteousness with healing in His wings, man enters the country above where visible angels and their judgments comrade with him. A new sunshine gives him new views of his fellow-man. No longer can he see the precious sons of Zion any other than the fine gold Jeremiah declared them to be. As the morning sun shows off the roses in the garden to be altogether different from what they looked to be in moonless night time, so people everywhere are judged differently by all who have crossed the bar into the bright upper Country.

Let us not look back to Chaldea, or Egypt, or India, for the great secret of miracle-working life. Let us look to the upper Country close at hand. "By strength of hand the Lord brought us out from this place," shall say the New Leaders heavenward. Let us spring past the bar of bondage into the free country of Health and Right Judgment. "Behold, now is the accepted time; behold, now is the day of salvation."

Caliph Ali was called the lion of God because high vision had taught him that man's own lot or portion in life is seeking after him.

"When ye pray, believe that ye have, and ye shall have," said Jesus. They all say the same things when they look to the upper Kingdom.

> "Far through the misty future,
>   Like an arrow of golden light,
> An hour of joy ye know not
>   Is winging its silent flight."

But how can we enter the joyous realm while we are bemoaning our lower lot? Even Dante saw that we must take our eye off our circumstances and look up to the River of Heaven flowing over our heads. He agreed with the son of Amoz: "And when ye see this your heart shall rejoice, and your bones shall flourish like an herb."

The Moslem knows that *knowing* that his own is this minute looking toward him is a drawstring on its hurrying toward him. Knowing is a drawstring:

> "Deep in the heart of thee,
> Soundlessly low—
> The Vach language wordless—
> Thy Soul pointeth, Lo!"

The gnomon in the heart is ever tending toward the upper Country.

"Homeward is the Tao's course."

By the law of Number Ten we learn to take back to ourself the lordship we gave to defeat, deafness, death. We take back to ourself the power we gave to people to hurt our feelings. We take back to ourself the power we gave to poverty, inconsequence, ignorance. We learn that the four and twenty elders of Paradise are gratified at our temerity, and shout, "We give thee thanks, Lord God Almighty . . . because thou hast taken to thee thy great power, and hast reigned."

By the law of Number Eleven we learn to use our hither drawn plenteousness of power to spring upward into our native and rightful New Wisdom, New Health, New Views of Life. The Egyptians saw in us, so re-empowered, a likeness to the Scarab drawing sustenance from the very dirt ball that humiliated it, and using that sustenance so drawn, to strenghten its down pressed wings to fly into free sunshine, free airs, free rains; free denizens of higher life. This is not only euphony, sweet wordings; it is humanly practical. Immediately people do stop hurting our feelings and sink into flatness when we take to ourself the power we gave them to hurt us. Immediately ignorance lets go and genius of a new order is ours when we take back to ourself the harming power we gave to ignorance. So our genius for rising as Fore-Helpers to the discouraged and defeated in the battles of human beings with their human dirt balls, fulfills the prophecies of three thousand years ago: "I will lead them in paths they have not known." "They shall mount up with wings, as eagles." They shall be an angelic ministry.

It is no wonder that Jesus of Nazareth could go in and out of the kingdom at His will, seeing that He had taken back to Himself the crushing power He had given to earth, and the hiding power He had given to heaven! How triumphant His voice: "All power is given unto me, in heaven and in earth." "Where I am, there ye may be also."

Was not Paul wonderful to say, "I live; yet not I, but Christ liveth in me." Were not the Brahmins wonderful to discover, "Brahma is you yourself?" Was not Jesus wonderful to explain, "Did not he that made that which is without, make that which is within also?" Himself always crossing the bar beyond the shadow system, forever finding that the Highest Self and the Inmost Self is one Self?

Men act foolishly by reason of judging from the evidence of the shadow system this side the Highest. As, if our money is snatched away we judge that we are deprived and humilated. That is judging within the shadow system this side the High Rich Zone. Looking up and crossing the judgment line we lay hold of the Hand that stretching hitherward giveth liberally Miracles of unexpected combinations, shadows of Finished Splendors from across the Bar do surely then transact in our behalf! So we judge another way and word concerning life. We judge with Abraham that truly "the Judge of all the earth doeth right." We jubilantly talk face to face, "In Thine hand it is to make great and to give strength." "Riches and honor come of Thee."

Wise Hebrews of old were watching for the power of the Highest to overshadow some divinity-sensitive human and drop a Messiah into the world. It is declared by certain Hebrews that many Jews have really believed that Jesus and Mary fulfilled the prophecy, though they were pledged to act against their own believing. As there is an exoneural influence from secret believing, we now see over six hundred million declared Christians as outward marks of the secretly held believing of the Hebrews.

For unexpressed believings come to outward showings. They are the idle words that burst forth into active exhibitions. They are still believings that find their way into loud results.

"That every idle word that men shall speak, they shall give account thereof." "Now is come the time of the dead (the still) that they should be judged."

An Emir of Cashmere had an occult reader of secret, unexpressed believings. It was his genius to lay his finger on the unconsciously held believings of the high court officials and prophesy exactly how they would actualize. Many an important personage was astonished at being removed from office. Many a lesser personage was amazed at being promoted.

The Emir sent the occultist to England to put his finger on the subsecret of England's greatness. "Britannia rules the waves, and the sun never sets on her dominions." This for the outward grandeur of a little island people must be explained to the Emir. He waited three years while the occultist watched for the hidden springs. Finally the occultist found the deepmost national secret of England's greatness. It was I. H. S.—Jesu Hominem Salvator—or, In Hoc Signo. (Jesu the Saviour of mankind, or, In this sign conquer.)

The Emir ordered the mystical formula I. H. S. to be inscribed on all his own possessions; curtains, rugs, gold and silver plate. "Go to, now, I also will be a great power," he said.

The Hebrews of old had an almost forgotten formula of faith. It was, "There shall no man be able to stand before thee"—I will contend with him that contendeth with thee."

On this account the Jews were not allowed to number their army. They might have an army; but small or great in number, the Augment thereof should always be the Unseen Almighty, in Whose victorious alliance they firmly believed.

They just as firmly believed that if they numbered their army it would be defeated.

David had always conquered all foes with his invincible soldiery. He had become proud and hard-headed as victorious kings always become, and numbered his forces, though Joab his leading general implored him not to number them, saying, "The Lord make his people an hundred times so many more as they be." But David made out the tally of his ranks, oblivious of Joab's protests.

Then Gad, who was David's occult reader of the outrushings of secretly held unspoken believings, told King David of the opening he had made for the outrush of the sub-reservoir of universal Hebrew certainty that numbering the army was disastrous. He showed David that numbering the army was doubting the Ally.

Within three days 70,000 of David's most stalwart warriors had fallen dead from pestilence. The king had made an opportunity for sub-believing to come forth with unrestrainable energy into nation-wide manifestation.

Jerusalem (the Self) is besieged to the eleventh year of Zedekiah, or Justice of Jehovah. Until the only secret believing is the boldly declared original truth of my Father's planting—the *Gnosis* not subject to defeat, there is ever the menace of some idle word. "Whatsoever a man soweth, that shall he also reap." "One jot or one tittle shall in no wise pass from the law, till all be fulfilled."

There have been two grandly uttered great believings in the history of man. They have sprung boldly forth from the divinely planted deep wisdoms of the inward parts. High Watch has struck back to them, and great have been the outward demonstrations of these two believings. Mankind has lately crossed them over with disbelievings, and sodded them down with refusings, till they are lost to view as original truth capable of mighty executiveness. They are sung and recited now long since as beautiful figures of speech, symbolics of unknown laws; but when God put wisdom in the inward parts, and showed the deep findings of upward look toward His Vast, Vast Countenance, He made these believings the energy of executiveness to save from war, pestilence, disease, death.

The first is believing in the power of His Revealed Name. "For this cause have I raised thee up—that my name may be declared throughout all the earth." "The name of the God of Jacob defend thee." "And this is his commandment, that we should believe on the name of his Son, Jesus Christ." "Whosoever shall call on the name of the Lord shall be delivered." "Our redeemer, thy name is from Everlasting."

The second deeply planted *Gnosis,* is that mankind does not fight for the Almighty; the Almighty fights for mankind: "Set yourselves, stand ye still, and see the salvation which the Lord will work for you; for the Lord shall fight for you, and ye shall hold your peace." "My kingdom is not of this world: if my kingdom were of this world, then would my servants fight." "Fear not, I will help thee, thou Jacob, and ye few men of Israel." "Not by might, nor by power, but by my spirit, saith the Lord of hosts."

One attempted believing there has been, over which mankind has stumbled and fallen and quarrelled much. Nobody has given mankind the clue, for nothing of it is to be revealed till the fulfilled moment of steadfast obedience to the ever uttering mandate, "Seek ye my face evermore."

This attempted believing has been as to the substance and nature of the Presence of Deity in the universe. Mankind has partly believed in the Being of Deity as Principle, and partly believed in the Being of Deity as Person.

As "Principle" demands reasoning, it follows that the ill-made or weak-brained are left out of the scheme of salvation. The weak-brained can hardly come at the albegra of "One Presence in the Universe; therefore as I AM, I am that ONE." To the very cleverest brained Jesus said, "Why reason ye?" as though the reasoning of the reasoners was not the light of salvation. "What I say unto you, I say unto all, Watch," He said; and He repeated the injunction.

There is no child so stupid but can be made to look

up to One ever beholding him, till gleams of intelligence steal down the track of his upward looking.

As "Person" intimates form and collect of parts, the term Person applied to Deity has stirred wide human resentment. "Who is like unto thee, glorious, fearful, doing wonders?" sang the children of Israel. "Touching the Almighty, we cannot find him out—with him is terrible majesty," cried Elihu the Buzite. Yet the children of Israel and Elihu the Buzite had come nearer to knowing Deity than any on earth before them, or, saving the Apostolic hierarchy, even since their inspirations. "The nature of Deity is undetermined," reads the most modern of Cyclopedias.

By the instruction of the Tenth Study in Divine Law we find that lower attentions strike back no farther than the shadow system gathered round the throne place of our inward being, while attention toward the Countenance of the High Redeemer inhabiting Eternity strikes back to the original Believer, the divine Gnosis, the throne place of our own being, where the knowing of the Lord Self lies deep.

Daniel got caught in a lions' den by much knowledge of cause and effect in wickedness. He was in the teeth of the law of what a man soweth that shall he also reap. Had he not aforetime thrown other men to the lions? But he had never let go the bold statement of his inward certainty that God should send His angels to keep him. "O King, live forever," he said, "My God hath sent his angel, and hath shut the lions' mouths, that they have not hurt me."

Likewise also Paul got caught by stones till he died of stones. Had he not in times past stoned other men to death? But his inward certainty that "He that believeth in me, though he were dead, yet shall he live," waked him from death. Thus mighty is an inward certainty firmly maintained, never ignored, never forgotten!

Nations with nation-wide certainties should remember them often, keep them flaming by bold ofttime expression. If there is certainty of conquering and salvation in *Jesu Hominem Salvator,* a nation should never talk or write of sacrificings of its men, but only of conquerings and victories by its men daringly going forth in the power of Jesu.

If it has once stated its trust in the Almighty to deliver with bloodless victory, it should never let the trust get cooled over by fear of armed foe or dread of failure of its Backer and Ally. On account of not fearing, wrote Isaiah, "Thou shalt be far from oppression."

"The eleven stars did obeisance to me" (Joseph). All the stars and the laws of the stars give way to the miraculous touch of the Supernal. "Joseph" means, He will add. He, the Supernal Unseen, will add to life as it now stands, life by the miracle. He will add to powers as they now show forth, powers unaccountable. Every attempt was made to slay Joseph, but the Unslayable Supernal was his shield. Poverty and slavery tried to crush him, but the Unseen Lord Uncrushable folded him as a buoying sheath. He was not afraid because he had made the Fearless One his high watch tower. Is not fearlessness a defence? But how does a coward generate fearlessness? By no process we may be sure except by much attention to the Author of fearlessness. Do we not show forth that which constantly attracts our inward gaze? Consider how the sight of the angels of Mons altered the looks and the conduct of the soldiers who beheld them!

"In the eleventh year—was the house finished," and Solomon said that all who looked to the house should feel its Mystical Rulership for miraculous liberty.

"House" signifies character. "Finished house" is character at its eleventh state, with all its mystical powers in full action. To look at the light burning in the upper chamber of the house of Rai Shalligram, postmaster general of North Western India, was to sense the effect of Rai Shalligram's Mystical Rulership. He had faced the splendor of invisible Brahma till his whole being was infused with the miracle of Brahma. He was certain that if he were naked and alone in a desert, Almighty Brahma would feed him, clothe him, shelter him, protect him from ravenous beasts. The rulership of his unshakable certainty was like far gleaming points irresistible to the men who caught sight even of his house. They all felt like throwing aside their families, their business, their clothes, their previous minds, and fleeing to trackless desert places to be alone with Great Brahma.

The "house" of Gideon had reached its eleventh stage. "Look on me" was all he had to say to cause Israel's armies to wake victorious liberty for all Israel.

Peter and John had touched the eleventh stage of character. 'Look on us" was all they had to say for natural laws to unclamp.

Character is judgment, and judgment is character. We are known to our neighbors according to our judgment.

If we judge according to Universal Protection, seeing all mankind embraced in Its mighty arms, upbearing, defending, as wings of eagle mothers under untried eaglet pinions, we give all who look toward our house new emotions, new judgments concerning Universal Allah, Brahma, God, as though for them a new light had broken on life and its mysteries. They catch bold assurance of Omnipotent Augment. "Their judgment and their dignity shall proceed of themselves," wrote the scientific prophet Habakkuk. Did not Socrates say that right judgment straightens out conduct, and wrong judgment deflects it? Did he not urge people to get right judgment by some attention to that which would eventuate in right judgment?

The establishment of the Jesus Christ judgment in man is the fruition toward which all religious and mystical laws are bent. "The Father—hath committed all judgment unto the Son," He said. "My judgment is just." The judgment day is come when all men judge the Jesus Christ judgment. It is the day of fire and brimstone to all other judgments. Appearances count for nothing. What stands back of appearances in its unalterable excellence holds undiverted attention. Is not man, back of his appearance of folly and decay an immortal being, stately, wise, flawless?

Ezekiel says we have been in Eden the garden of God, when we have looked toward our own deep rock integrity, our hidden God Self: "Thou hast been in Eden, the garden of God, and the ruby is thy covering." Taking Martin Luther's advice to let the Scriptures interpret themselves, we let Job interpret "covering" or robe. "My judgment was as a robe and a diadem," he declared. So the ruby signifies right judgment and kingship. "A king that sitteth in the throne of judgment, scattereth away all evil with his eyes."

No mistake is made by one who looks straight forth from his integrity centre. He sees a full barrel of meal where the widow sees bare boards. He sees genius where others are seeing failure, or death, or insanity. He is the judge who sees the way to raise the dead as the mathematician sees the way to cube the fraction. He sees the right relation of man to man, and his word is the expression of his sight. Did not the word of the Lord in Elijah's mouth raise the Zarephath child? How had his word caught the spark of such executiveness? By his recognition of his own, "I, the man of God." How had he come to such high estimate of his own "I"? By much attention to the Lord God of Israel, before Whom he declared himself as forever standing.

The *Eleventh Lesson* in *Divine Law* tells of the integrity within as right judgment resident in all men alike; covered in all men by flesh and its mind as in a crypt of darkness. As silent is right judgment as if dead. It is called the mountain of Zion by Jeremiah, who wails that the "mountain of Zion which is desolate, the foxes walk upon it." He means that small, low, mean, myopic judgments run ahead and formulate our speech: "The precious sons of Zion," he said, "comparable to fine gold, how are they esteemed as earthen pitchers."

As Elijah's recognition of his "I" as the word nigh him, even in his heart and in his mouth, came from much recognition of the High Redeemer inhabiting Eternity, and this looked forth from him as right estimates, so other men catch their self recognitions from associating with other stronger men, and pass on their judgments wherever they go.

"Many seek the ruler's face, but every man's (final) judgment cometh from the Lord."

The Ganges mother loved her baby. She hugged it to her with passionate devotion. But she threw it to the crocodiles, because her husband's judgment was the ruler's face that she was daily seeking. So her judgment catching fire with her ruler's judgment struck forth into action against her beloved baby.

As bamboo stick striking bamboo stick sparks fire, so judgment sharpeneth judgment.

Notice the difference between the prophet Elisha's judgment striking against the Shunamite's inner chords, and the judgment of the Ganges husband on the Ganges mother's chords. Elisha was judging always the judgment of life, joy, victory. "It is well with thee?" he asked the Shunamite, whose heart was breaking. So strong was Elisha's living agreement with life and joy that she answered, "It is well." "Is it well with thy husband?" asked Elisha. "It is well," she answered. "Is it well with the child?" "It is well," was still her word. The smite of Elisha's judgment on the waiting chords of her being struck forth the spark-speech that waked life in the dead child, joy in the husband, the day of judgment in the mother.

Saul caught fire with the judgments of the prophet Samuel by constant association, till the men of Israel were astonished at Saul's prophecies.

There is One with Whom we can associate till our "speech bewrayeth us." His Vast, Vast Countenance shining as the sun, now beams upon our heads divine wisdoms, and glows life-giving actions in our undertakings. He is the Great Ruler Whose judgments strike fire with our waiting chords, till the Lost Word sparks on earth, with its instantaneous miracle-working energy. The life coal lingering in the faint must fan to flaming fires of undying Omnipotence at the smite of the Lost Word on the ethers. The joy chord mute in the breast of humanity must tremble with good news as from a far country at the Lost Word's sweet import.

The judgment of the Great Ruler is on the Finished Estate that faces us from every infinitesimal point in the universe. All things and all people are adjustable at every instant to their own finished estate of flawless excellence as judged by the Great Ruler's judgment. They all wait the promised New Speech to show their own finished fact.

"Give ye ear, my judgment is toward you," saith the Great Judge. "And I will come near to you to judgment."

"Pray as if you had already received," taught the Master. Of course we have already received, if it is already finished and near at hand. Is it not waiting the lighted eye to make the prayer of acknowledgment a simple truth?

The angle of repose in physics is that angle at which one body may rest upon another without falling. The angle of repose between God and man is where man's judgment and the judgment of God agree. "I have put my judgment in thee for a light—Here is my rest forever; here will I dwell," saith the Judge of all the earth.

"They neither marry nor are given in marriage," said Jesus. Of course not. They are already married from all eternity under the eye of Allah. What is all this marrying and giving in marriage but trying to adjust to the Finished Marriage? This great truth was visible enough to "Baal Shem" so that he could sometimes see which people were married, and he led them across the breadth of the land to meet each other.

He knew, like all advanced Moslems, that What Ought to be Is. "If we ought to be there, then we are there; let us sleep." And a week's journey was performed in one night, while they slept, because the great Moslem rested on the *Ought* to be that *Is!*

Solomon was the forerunner of the resters in the *Ought* to be that *Is.* He knew that according to the judgment of Jehovah God he already knew, therefore he knew. And though he had not studied them he told of the stars and the stones, and of the growing things, from the cedars of Lebanon to the hyssop that springeth on the wall. Jesus knew that according to God the Father He already knew all things, therefore He needed not that any man should teach Him. "Is it not written in your law—ye are gods," He asked the people.

As there is no respect of persons with the Highest I AM, He told His disciples to call no man upon the earth their Father, regardless of appearances, that the straight line of healing energy from their Father might run through them, to the revealing of what verily *Is,* everywhere they walked. Thus to them the apparently dead were as alive as the apparently living. To them the sick were only as straight sticks deflected in water. So they took the sick by the hand, and pulled them standing. "Rise and walk," they said.

They were hard masters, reaping where they had not sown, for they had not to sow the seeds of good instruction and see them spring up where the sick or dead were concerned. They saw that what already *Is* was under everybody's feet, and all anybody had to do was to stand on it!

Right Place. Right People. Right Powers. Right Possessions.

This is the High Science of Jesus Christ: "Go thy way, thy son liveth."

Anthony of Padua suddenly saw into the kingdom. From his pulpit in Padua he heard them telling the woman that her son was dead. "He is alive!" shouted Anthony. And to the astonishment of all Padua, not one of whom had any knowledge of the Unalterable, Finished Fact, the boy they had seen dead was found to be alive.

The Finnish woman subconsciously knew that according to the judgment of the Lord Inexorable she could not be in debt, for what was hers was hers unalterably, and what was the other's was his from the beginning, and nothing could take from or add to the original endowment. With some awe but not with any astonishment, she saw sixty dollars multiply to one hundred and sixty on the table under her fingers. As she was only reputed to be in debt one hundred and fifty, she wondered for a moment what the extra ten might be there for, when she suddenly remembered that she had always believed the till would never be empty. Her forgotten believing had come to open showing. The word, idle even as the dead, had come to expression. "Now is come the time of the dead," when some heaven-born recognition of the Eternal Actual wakes in mankind!

"And the *Eleventh* was the jacinth." Pushing the red of the jacinth to its highest value we have the ruby, symbol of beauty and symbol of judgment. Beauty and judgment present the same credentials for our favor. They represent the proper adjustment of part to part, man to man. A good judge is more sought after than a king. He puts man into right relation with his fellow-man.

Beauty is the proper adjustment of part to part. "How beautiful—how beautiful!" we exclaim of the Taj Mahal of Agra, India, and of the Mosque of Omar at Jerusalem. Yet each part of each temple is no more beautiful by itself than the similar parts of millions of other buildings. It is the proportion, the balance, the inimitable adjustment of part to part of the Taj Mahal and the Mosque of Omar which satisfies our secret Zion standard.

"The Lord shall send down judgment with fire." "Thou shalt not pervert the judgment of the stranger." The greatest stranger to man is God the Lord. Some heard the sound of His voice and said, "It thundered." Others said, "An Angel spake." But Jesus heard the voice of God in coherent speech: "I have both glorified it (My Name), and will glorify it again." The judgment of the Stranger was perverted by the people, but Jesus received it as the mystery of that Name which when called upon saves from sword, famine and defeat. He sealed to mankind the foretellings of the great prophets, that His Ineffable Name, key to the mysteries of the universe, should part the ages old silence with the promised new language, hastening over the Tao or Track of High Recognition.

"And the *Eleventh* lot came forth to Eliashib"—whom God restores. "I will restore to you the years that the locust hath eaten . . . and the caterpillar." "If any man sue thee at the law, and take away thy coat, let him have thy cloak also." For what is the man suing? Surely he is violently striking out for adjustment to his own lot or portion in life. He strikes for judgment. He must not mistake his man. Meet him with the judgment firing one who has companioned with the Owner of the spheres. Give him the goods he strikes for, till by thee he also companions with the Original Giver and Divine Restorer to each man of all his inalienable rights. "With what measure ye mete, it shall be measured to you again." There are coats and cloaks of greater value, to take the places of the coats and cloaks wrested from man by the other man on his rough way to his own goods.

"Whosoever shall smite thee on thy right cheek, turn to him the other also." He, as before, is on his violent, mistaken strike for his own good. Let him not mistake his man. Nothing shall by any means hurt thee as the man strikes fire with thy right judgment: "No man taketh my life from me." So estimating the blows of human encounter, we cross the bar between earth and heaven, and the ruby blood in our veins is the morn-glow of the New Kingdom. "I saw the city," said John; and boiling oil and slander felt the estimates of the John Supernal. They had no power to hurt.

"Thou shalt not steal." Is this a commandment or a promise? This must be according to your ruler's face, law maker of the Ganges, or Lord Jehovah of the Fixed Estate. One veils the sight of the Inalienable Undiminishable that faces us. One exposes the Self Existent rounded Good not to be added to by action or non action forever.

One whose judgment has been the strike on our secret chords causes us to shut our neighbor away as a diminisher. One causes us to see him driving for our judgment's bamboo smite, passed on from the Lord of the Inexorably Undivided, who sees the everywhere fixed lot awaiting its owner's acknowledgment. Your speech "bewrayeth" your class.

"Thou shalt not take the name of the Lord thy God

in vain." Have we not discovered that the Name of the High Redeemer is the Lost Word, the Apocalyptic trumpet, which causes the shadows to flee away from the dead, the sick, the unhappy, the earth hindered, the instant it is spoken? *It cannot be taken in vain!* It is instantaneous in its rending of the curtains of hiding. Has the name *God* this instantaneous sign of its executiveness? What name can we not take vainly? We have learned that as the language of our lips "bewrays" our associations, either with the learning of the schools, or the Knower to the schools unknown, so our answer to the Mosaic decalogue "bewrays" our God.

And our acquaintance with Jesus Christ, Holder of the New Name that cannot be taken in vain, tells whether we are nearing the speaking of the Name that cannot be taken in vain, or driving round about to find it. For the Lord Christ only hath daringly declared, "I will give a New Name, which no man knoweth, save he that receiveth it."

"Those that thou gavest me," said Jesus. "I have kept, and none of them is lost save the son of perdition." "Son" is idea, and "perdition" is loss. I have lost only the idea of loss, He said. The idea of loss is Judas. He is lost; and in the place of Judas is Matthias, the gift of God.

And Matthias is numbered with the eleven." He is the value of all the eleven in one, as the gift of Elohim compasses all the good mankind could ask for, or even think, as good.

Is there not one sense of loss that haunts the best of the sons of earth? The death of that one sense of loss is the death by which he must glorify the Great Restorer. In the river Euphrates, or the river of human life that flows through mankind, we dam the onflow of original goods as they drive from the banks of the Land Beautiful to the Port Supernal. We dam by hugging hard to goods that harbor the Judas miasma of loss.

Approbativeness is hugged with its Judas miasma of loss. The dog whines under the table when its neighbor dog is praised. Yet no praises bestowed on the neighbor dog detract from the whining dog's standing. It is only the idea or sense of detraction that hurts. Could the unhappy beast know the truth he would be free from his miserable Judas.

This is the death by which many, even great people, might glorify their Heavenly Giver's impartial munificence; namely, by letting others be praised to high water mark for what they themselves, the great people, feel they deserve all recognition.

They must stand to the truth which the dog cannot appreciate. They must know that the praises belonging to them are theirs, and are near at hand; nearer for the spaces made by the departing encomiums. "I will also glorify them, and they shall not be small," saith their Judge, their Great Original.

Affection is hugged with its Judas sting of the possibility of loss. What a log it is among the goods driving down the human life river! Somebody wins away our beloved in whom we have delighted. No hope remains. Philosophy may declare that when half gods go the gods arrive, but neither philosophy nor good will can help the heart to let go its one only among all others.

But why should we not trust to the Voice of the Highest? We are not dogs that cannot listen to the voice of Revelation. If the beloved is ours, none can remove the beloved! "I will restore," saith the miracle-working Restorer. Confidence in the miracle of Jehovah is the ruby stone among the gems of character. It colors the blood with living fire. Let go, for the miracle of Jehovah! Make way for the miracle!

Ambition is charged with the miasma of humiliating Judas. How many of mankind have dropped out of sight, because their high seats have failed them that others might succeed in their places. It is the agony of kings, emperors, presidents, officials all along the lines of human emulation. Who has had voice to reach them with the divine "verily, verily," that seats, thrones, honors, let go when they let go, making way for the Restorer's miracle of honors? Age nor sex nor handicap can hinder the coming of the miracle of the Unseen Giver. The high seat belonging to each of earth's millions on millions must see the old ambition log in the Euphrates life river give way, before it can come into place. "Sit thou in that seat, and none can make thee rise," declareth Allah the Inexorable.

Acquisitiveness is another sign of Judas. The fear of loss of possessions goes ever with the acquisition of possessions. Nothing can neutralize the secret fear of loss in the breast of the owner of goods, save attention toward the mystic light of the Inexorably Determined that shines unnoticed over his head. What Allah has given cannot be diminished. And Allah has given more than man has amassed or ever shall amass!

Man's face turning toward destruction, how can man help judging according to destruction? We judge according to our attentions.

The watcher toward Unlimited Ownership in the Upper Realm so near, has inward assurance of security.

Why do men build battleships, or lock their doors? Surely it is because of fear of loss of some sort. Nothing voids of the Judas fear of loss till high watch teaches free letting go of fleeing goods for the coming of Matthias, the gift of the Father Munificent.

Letting go, letting go! How free may be the life river when we know the wherefores of our denudings! "This he spake, signifying by what death he should glorify God."

"Wheresoever thou lookest, there is Allah."

From every point in the universe the question faces us, "Is it well with thee?" And to every point in the universe the question-smitten chords of our being answer, "According to thy judgment it is well with me; therefore it is well." By so agreeing, man strikes beyond the pairs of opposites, success and ill success, honor and dishonor, and is he of whom the Apocalyptic seer writes, "To him that overcometh, will I grant to sit with me in my throne."

The vision of the Universal Judge ever toward the sons of men, being met by the return gaze of the sons of men, drops into the beautiful speech of the New Age just at our gates.

The Beatitudes of the Great Judge are all uttered as invigorating high science to the initiated. Notice how the *non initiate* stumbles at the blessedness of being reviled and persecuted and spoken evil against falsely. But the *Initiate* is no crying saint or rebellious Nietzsche. He knows that the persecutor by evil speakings and hateful revilings is knocking at the right gates. The Initiate at his judgment seat stands to it that the reviler is only violently taking the kingdom of his own heaven. The evil speaker is striking for right judgment. He is the Shunamite over again, meeting his Elisha the Tishbite, who does not flinch from yielding the basic good that the reviler is earning by hard labor, though he knows that the reviler might have had his good by an easier and quicker method.

Elisha did not quail at the signs of weeping. He held his own. Let us never flinch at what is happening around us. What is transpiring above us is our concern. And this wins all battles; for the kingdom across the bar is the Ruling Kingdom. It is the high watcher's business to get all the world to be interested in the Ruling Kingdom for which its heart is truly yearning.

The Judge that standeth up in the universe facing us, saith, as if testing us, "Is it well with thee?" We answer "According to thy judgment, it is well with me; therefore it is well." We find that this bold communion accomplishes a strange victoriousness over all our undertakings. Being well with us according to high judgment, we see that it is well with all others. Over all habitations we see the Other Kingdom glowing. We hail complaining speech as divine challenge; our great opportunity. Under all feet we are aware a finished globe rolls with wonderful offerings. Let us salute the ever waiting answers to our prayers. Is it not declared, "Believe that ye have, and ye shall have?" Is it not declared, "Thy lot or portion in life is seeking after thee?" I counsel you to heed no contrary clamorings. "Who is deaf as the messenger I sent," saith Jehovah Nissi. "Salute no man (or opposition to The Judge) by the way." The long looked for Judgment Day is within our gates. Let us joy in the fire of its purifying. Let us catch its brimstone Truth, the mysterious, great and mighty things not yet known to this world, which the prophets all heard announced as coming in with High Watch and glad accord with High Judgment.

The Sacred Books of all ages mention three sciences: Material, Mental, Mystical. Material Science declares laws that are sure; as that iron sharpeneth iron, and hydrogen and oxygen clashing together fall into thirst-quenching waters.

The Sacred Books proclaim a Mental Science to which the world can subscribe, as, "All that we are is made up of our thought."

Mystical Science announces the miracles of "Predicateless Being," setting the ways of matter at naught, and nullifying the thoughts of mind:

"The flesh profiteth nothing."

"Take no thought."

"In such an hour as ye think not."

Mystical Science is a chalice of golden wine passed along to the sons of men by John's angels of the Apocalypse.

It is a new song for the hearts of the Children of the New Age.

E. C. H.

XII

MINISTRY

— a —

The twelve gods of the Egyptians were twelve rulers of this world declaring twelve obligations for this world to fulfill. The astronomically prophetic Pyramid of Gizeh was built by the duo-decimal system which finally arrived as the twelve inches of the English foot, every inch alive with its own god law. Twelve is the number of Mount Zion: the number of the Heavenly Jerusalem.

The twelfth stone is the amethyst symbolic of Revelation of the Mysteries; discovery of secret values; the hitherto valueless coming to light as long hidden Royal Passports.

Do we highly value idleness? Yet the idle hold the key to the city of the Great King. "Labor not," said Jesus. "Your heavenly Father worketh." Therefore, why should they work? And the careless also, whom we hold in light esteem; yet the high law reads, "Cast all your care"—"Your heavenly Father careth"—Why should you care? The thoughtless, the thinkless, we will not praise them, but they hold the heavenly key of "Take no thought"—"In such an hour as ye think not"—"Your (heavenly) Father knoweth." You do not need to think.

A new accomplishment, a new protection, a new knowing, shall come to the user of this hidden key that opens golden gates to walls of Zion all jubilant with song.

Another feared and half despised condition is old age. How afraid people are of old age, covering all its symptoms with powders and paints and dyes to show how they hate it as an undesirable! Who has told the aged that the secret of a forgotten science hides under faltering brain and stiffening limbs? Who has told them to cease from noticing the faltering and the stiffening and regard with new regarding the science of sparkling re-living hiding under these, ready to break forth with power of youth such as no youth has ever known?

Now and then along the centuries some old, some very old, person has stirred with the hidden science and has done what no young person on all the earth could do, or seeing it done could duplicate.

Joshua at 85 years of age quickens with a mystic tone that throws down the walls of Jericho and starts the Boaz family of forebears of royal David, forefather of *Jesus Nazarenus Rex Judaeorum.*

Moses at 120 years of age inspires with heavenly fervor that shakes the still Jehovah wisdom in Joshua as he lays his hands alive with power upon him. No younger man than Moses at 120 could wake the Spirit of wisdom in old Joshua so that he had almost risen into the heavenly Jerusalem in supernal manifestation among the great ones of Israel.

John at Rome in the cauldron of Domitianus, at 80 years of age sets the boiling oil at naught by some mystic atmospheric so generated by the Sacred Name that the Apostles at Ephesus carried him into church with a plate of gold on his brow on which was inscribed the Name of majesty he had so proved as able to save to the uttermost. No young man had come out of boiling oils alive. Only John at 80. And at 89 he, still holding his mystic atmosphere, was most honored among the Christians of that New Age just sending its first golden beams over this dark earth.

Massini in Firenze singing Gounod's Sanctus at 70 years of age so entranced the people that some felt themselves transported as to a Paradise they had left behind them or a Paradise they were moving toward. No young singer in all the choirs of Italy could touch the still chords of love, rest, and home like old Massini.

It is for us of the age just shedding its promised daylight over our earth to wake again the slumbering science of re-living so that something diviner than youth may fling its celestial signals forth from behind the falterings of ages-old human processes.

Man miraculously victorious shall stop his persistent reiterations of descriptions of Deity and His One Idea Man, and urge face to face recognition of Deity as Responsive Servitor.

"To him that ordereth his conversation aright will I show the salvation of God."

Learn to converse face to face with The Ever-Facing All-Knowing.

"Therefore, behold, I will proceed to do a marvelous work among this people." "I will do a new thing; now it shall spring forth!"

"And there is no discharge in that war," spake Solomon, catching in one of his high moments, the mystery of the amethyst stone holding mystic signification as endless inspirings from endless Wisdom.

Mary of Bethany caught the light of the Endless and started the Christian Church numbering millions on millions of people, and endlessly increasing millions.

"Other foundation can no man lay." These are wonderful words. They stand for the power of a Name. They stand for the mystical influence that wakes the heavenly power called Inspiration, Holy Ghost, working miracles upon this earth, making new conditions like in beauty and joyousness to that other realm called by the Brahmins the Perfect Land.

Daniel, royal captive to the King of Babylon, 500 B. C., felt the Spirit, or inspiration of "the holy gods," and three kings set him into high authority as wiser than all the wonderful wise men of old Babylonia. They found him charged to the brim with mystical knowledge. Daniel knew a name which woke inspiration: "Blessed be the name of God forever and ever, for wisdom and might are his," he said. "He revealeth the deep and secret things."

So today we come again upon the heavenly mysteries all known to Jesus of Nazareth, royal world captive, who offers to the world at large what Daniel offered to the three dynasties, Chaldean, Median, Persian, more than two thousand years ago:

"In my name he (The Holy Ghost) shall teach you." The Holy Ghost hath a Voice. John turned to see the wooing Voice. Now and then one of us hears the same Voice. Sometimes it is a soundless chord in music, like what Ole Bull caused whole audiences to be stirred by, after he had roused their inner ears beyond the power of his bow across the strings.

Whoever shall continue to hear the unrecorded strain shall fetch forth the prophesied New Music. Sometimes it is in apperception, as great teachings smite our heart chords. Whoever keeps the apperception shall give us the Vision of The Presence that sweeps this world aside for the New Kingdom to be our dwelling place.

The teaching and miracle-working Holy Ghost saved Daniel in the midst of the starving lions. It saved the three wise men in the midst of the fiery furnace. It saved Paul stoned to death at Lystra.

Daniel was in the teeth of the law of cause and effect by reason of having thrown other men to the lions. He was in the law of "What a man soweth, that shall he also reap." But he having touched a law above human cause and effect came forth unharmed.

The three wise men were in the teeth of the law of cause and effect by reason of having thrown other men into fiery furnaces. They, scenting the winds of the other world, set aside the law of cause and effect and stood forth unharmed.

Paul who by the law of sowing what we reap was stoned to death as he had stoned others to death, having acknowledged the New King, rose up alive, happy in the Living Kingdom as the Sibyls had foretold. Death lifted its cloud and passed away as the Sibyls had foreseen; the Cumaean Sibyl, the Erythraic, the Samian—did they not all a thousand years before Paul's time declare that to acknowledge Him who should come as king was to rise free and happy in His kingdom, though "hostile men should spit upon that king, and on His sacred back they should strike"?

All the Mystical Laws declare Christ Jesus of prophecy, history, and the Undescribable New Age— this very age swept clean of delusions by the breath of Atman, showing our feet standing not on old earth but on a New Earth forgetting the former things as if they had never existed.

— b —

Great events of history, striking to the heart's core of mankind, are heralded by signs in the sky, and earthly eccentricities.

Was it not recorded that at the birth of Jesus of Nazareth the pole of the heaven stood motionless, while everything on earth which was propelled forward was intercepted, workmen with upturned faces and hands mysteriously suspended in air, and cattle strangely pausing at their fodder? Did not even the far away Chinese post it in their astronomical tables that a new star burst into the heavens at that date?

Has not the earth lately slowed up on its axis, the sun shaded its face with new dark spots of mystery, and a more splendid sun come gleaming forth in far distant night skies? Are we oblivious to the fact that these all have presaged the unexplained Armageddon of old Biblical prophecy?

Our globe has crossed a bar, and no prophecy based on previous astrological data can set us straight as to our status in our own constellation. We are driven to the visions of Daniel, Malachi and John the Revelator for sole comfort and edification as to the finals of the vast Apocalyptic combat in this twilight of the gods.

This is the time of which Joel and Paul were foretelling, when only those that call on the name of the High Deliverer shall be delivered from identification with the conflict in which all others on earth are occupied.

Mystical Science, top currenting the material and mental sciences that run through the prophetic Sacred Books, lays large stress on the hurrying assistance of the divine Name, whose sounded syllables catch up the flakes of living ether that lie in the common airs softly awaiting inbreathing recognition.

Mystical Science lays large stress on noticing the surrounding ether's soft healing flakes. It teaches to practice inbreathing the waiting Christ breath with its vivific stimulus. It urges to bide the time of the healing elixir's kindling all our flesh with the freshness of its own eternal fires. "For is he not Atman, the breath and the life of the universe?"

There is a vital difference between talking *about* the inspirations that wake the waiting God-Seed in the breast, and practicing healing inspirations as daily breath.

There is a vital difference between talking about the Judge of all the earth, and answering face to face the Judge's unceasing decision as questioning, "Is it well with thee?"

Answering face to face according to the Great Judge's inexorable decision is the mystical recognition that personifies. One comes into our life judging us according to our best only, as the great Judge judges; and so judging he warms forth our best into flowering beauty. One comes causing our hidden genius to bloom. One comes shoving us into our rightful environment and delivering to us our rightful possessions. One loves us for our native, original excellence.

And if one thus loves us and judges us, another comes thus loving and judging, and still another, and another, till we are surrounded by the personifications of our acknowledgments of the decisions of the ever-facing Judge of purer eyes than to behold iniquity.

If we have spoken face to face with the Invisible

Highest as Wise Counsellor, we shall find the personification of our acknowledgment in some new friend's
arrival all alive with high counsel.

If we have spoken face to face with the Invisible
Highest as Powerful Champion, some unexpectedly
greatly powerful champion of our cause hails into
comradeship.

For each acknowledgment face to face with the ever-
facing Mighty Judge personifies. Is it not promised, "Act
as though I were, and thou shalt know I Am"? And every
personification multiplies, till our whole field of life is
dotted with champions and counsellors and wise judges.
Therefore, endure "as seeing him who is invisible."

Endure to the personifyings. Catch the love fire of
the Sun of Recognition on the stalk and bud of hidden
possibility everywhere facing us. "Wheresoever thou
lookest, there is Allah."

The sunflower endures as seeing the sun though
the sun's glowing face is nightly obscured by a dark
globe's eight thousand miles of thickness. It squeezes
in the sun's hot beams with ecstatic adoration, till living
seeds fall from its yellow bosom and other sunflowers
spring up and fill the garden patch with sunflowers.

So did Gideon find his senses enthralled by the angel
of peace, Jehovah Shalom, till all the Ophrah field was
peace for forty years. So did Cosimo see peace, with all
his senses enthralled, till Firenze was at peace throughout her borders. So shall "thy land (be) Beulah, . . ."
married to the Lord," and great shall be the peace of thy
children"—sweet, effortless propagations!

In whatever time of mankind's history we read of his
steadfast attention toward the Vast, Vast Countenance
of the High Redeemer inhabiting Eternity, we find him

personifying his high descriptions with people like unto his descriptions. "Thou art my light and my salvation," sang the Hebrew captives in Babylon. Then came great Cyrus lighting up their half forgotten religion, and his Persian army with him, to hew for the singers their lost way back to Jerusalem their half forgotten home.

In our own time, by practicing the same law in lesser fashion, the young Dorman declared himself so identified with Omni-present Spirit, all-health, that like the ardent Sufi he felt "O Thou I and I Thou!" This made him as "the Sun of righteousness with healing in his wings." All who came near him sensed healing elixirs stirring in their veins, and blessed his miracle-working sunshine.

Dorman did not notice the people's wailings. He did not grieve for their misfortunes. He was deaf and dumb to all save the healing "Thou I and I Thou." "Who is blind, but my servant? or deaf, as my messenger that I sent? Who is blind as he that is perfect, and blind as the Lord's servant? Seeing many things, but thou observest not; opening the ears but he heareth not." Thus the blind and the deaf like the idle and the aged hold the mystic key to another world's wonderful gates. Are we not all of us open-eyed and open-eared to signs and cries of distress? But the truly open-eyed see the heavenly host free, wise, immortal.

The twelfth stone of character is the amethyst, significant of divine deafness and blindness. As the ruby is the most precious of all the precious stones, in the estimation of man, being significant of judgment, so the amethyst is least precious among men, being significant of Dorman deafness and blindness, two terrifying conditions in the estimation of man, depending for their supernal transactions upon face to face recognition of the transmuting God. "Though they cry in mine ears with a loud voice, yet will I not hear them," saith the Lord. Yet "I am the Lord that healeth thee."

Is the sun mindful whether it warms the rock or freezes the water? So on my hot hate the healing heights let fall a cooling silence, and on my cold despair they drop a loving warmth. Notice how the Great Unnoticing heals its opposites!

> "Now last of all comes number twelve,
>   And what should that recall?
> The Apostolic college
> When completed by Saint Paul."

Paul was the College, or collect, of the marriage of all the twelve Apostles to the Risen Christ Jesus, Victorious, Almighty. Never such lovers on earth! They were so in love with their Risen Lord that boiling oil had no terrors for them, lions' jaws no hurting menace.

Paul was the transmuted of the Apostolic lovers. He was transubstantiated from Saul the persecutor of the Christians into Paul chief star in the Christian firmament, founder of the Protestant Church, the most stalwart and vigorous religious body on earth. "I am the least of the Apostles," Paul said, ". . . because I persecuted the church of God"—"But God hath chosen the weak things of the world to confound the things which are mighty."

Thus was the beautiful dogma of transubstantiation, believed in by the steadfast Roman Catholics, demonstrated beyond cavil by Saul-Paul, and set in the earth as the mystic amethyst of Christianity.

The twelfth lot cometh forth to Hashabiah, twelfth

leader of the twelfth course of Levitical Singers. "Hashabiah" signifies such as "Jehovah esteems." "Hashabiah" is he who is set as a precious stone when the Lord maketh up his jewels. For the Hashabiah type sings in the night time of adversity and in the day time of prosperity; in the prisons of enemies and in the castles of friends.

Songs of the heart's Hashabiah well-springs bubble up through external affairs, send shrapnel of tribulation or thunders of applause. Paul and Silas in the dark Macedonian prison are singing with their feet in the stocks,

> "Send down thy Spirit free,
> Till wilderness and town
> One temple for thy worship be,
> Thy Spirit, Oh, send down."

And the inbreathing Free Spirit fills their prison house to bursting, breaking apart the binding stocks, and transmuting prisoners and prison keepers to worship of the Sender-down of Omnipotent Free Spirit.

"Thrice was I beaten with rods," said Paul; "once was I stoned; thrice I suffered shipwreck." "But I glory." And thus glorying, even Paul's hands had transubstantiating effects; for it came to pass that "when Paul had laid his hands upon them, the Holy Ghost came on them and they spake with tongues, and prophesied."

In the alchemist's cupel the precious metal was separated from the vile amid chemical violence. Thus was Elisha's anguish at the departure of his beloved Elijah turned into transmuting beams that cured brackish Jericho waters, by his crying out, not for departed Elijah, but for the God of Elijah: "Where is the God of Elijah?"

The sand grain or the insect's torture in the amethystine shell wakes the sleeping nacre in the oyster's cold bosom, and the priceless pearl is created. So "surely the wrath of man shall praise thee," sang the Hashabiah singers. Was not Jesus angry and grieved? Who is there among us quick-witted enough to transmute his rage into some glowing tribute to Jehovah's healing responsiveness by shouting to Him, "Stretch forth thy hand!" in imitation of the prompt-speaking Christ Jesus? Jesus thereby used the philosopher's transforming stone for which the magi vainly sought. He stirred the fabled health fountain lying deep in Him as in all men. He struck forth the elixir vitae of the God Presence, by shouting aloud His praiseful transmuting recognition, in the midst of his anger and grief.

Who uses the hot words that exactly express his anger or grief, and then is shocked at their reactions? Who gives excuses for terribly efficient exclamations that close the portals to miracle working? As well give excuses for angrily thrusting hands into molten metal!

Take lesson of Jesus of Nazareth, and in hot anger or violent grief use the portal opening words, "Because of Thee I am greater than whatever can happen to me! Because of Thee I am richer than any riches that can fall to my lot! Because of Thee I am more befriended than by the great friends that come to me!" Even on the cross, the Hebrew translation translates, "How thou hast glorified me!"

And the twelfth lot cometh forth to Jakim, "whom God sets up." Whom doth God set up? "I will set him on high because he hath known my name."

Jakim and Jacob being the same consonantally, and holding the same relation to names as mysterious energizings, we see how Jacob was the cord of his own inheritance, by the use of some name of that High Redeemer to whom in his lowly estate of timidity and sin he continually looked for miraculous helping.

The greatest blessing David could think of to confer upon his people, was "The name of the God of Jacob defend thee."

Jacob conferred names on his sons which had in the mystic potency of their sounded syllables the power to compel victorious outcomes. When he told Asher that he should dip his foot in oil, he meant that Asher by holding fast to the secret name of God folded within the syllables of his outer name, all his happy truths should prosper. Was it not a happy truth that the Sufi sang, "O Thou I and I Thou"? Was it not a happy truth that the captive Hebrews sang, "Thou art my salvation"? Was it not a happy truth that Dorman sang, "O Thou Free Spirit"?

They must have dropped down the line from Asher, the steadfast-to-the-secret doctrine held fast in some name they used, till it unlocked the mystery of quick successes.

When Jacob told Reuben that "unstable as water, thou shalt not excel," he meant that Reuben's secret name of the Highest Helper would hold him forever steady in the midst of his unstable tendencies, if he would steadfastly declare, "I, Reuben the steadfast!" Down Reuben's line, by his fidelity to the whispered secret name folded forever in his outer name Reuben son and friend of God, should spring forth a new order of great men to bless the world.

But Reuben never could remember, except here and there and now and then at odd intervals, to declare "I, Reuben the steadfast," till the secret saving name given

him by his father Jacob, the name within the name en-
larged him beyond his temper and his talents. Reuben
neglected the mystic principle as moderns have neglected
the mystic possibilities in the Jesus Christ syllables. For
the name *Jesus Christ* holds within its claspings the Lost
Word with its power to open the gates of heaven and
breathe through our human frames mysterious wafts of
immortality. Those who neglect or reject its offers are
those dropped down the line of Reuben first born of
Jacob and Leah, by inheritance not over-handicapped,
but by reason of downward viewings out-stripped and
out-done.

When the birthright of Reuben was taken from him
and passed on to Judah the ever-praiseful, the ever-re-
joicing seer of pillars of excellence standing up in hardships and handicaps, Jacob handed out to us one of the
mystical lessons for which he is now honored: "I, Judah,
recognizing the precious fire charging my own name as
precious fire charges all names, ready to spring forth
with new powers by recognition—"

Jacob was teaching the terrible alternative of moral
and mental and material life activities: *If* we stand on
our feet we can walk. If we do not stand on our feet we
cannot walk. If we know that our thoughts come from
our inward viewings we ofttime view Godward; if we
do not know this origin of thoughts we ofttime view
down and perforce must ofttime think low.

Was not this the ever-presenting bar of "terrible
alternative" mentioned by the ancients? Jacob showed it
by human beings, giving each son a name with func-
tions for triumph if held heavenward, but closing down
in unmiracled warfare and labor if held as man to man
or man to dust.

"For this cause have I made thee stand," said the voice of Jehovah to Moses; "that my name may be declared throughout all the earth." "From the rising of the sun shall he call upon my name," seeking My face evermore.

Attention to any objective makes duplicate, replica, in achieving powers as in character. "By me kings reign," saith the King of Kings and Lord of Lords, Ruler in the heavens and the earth.

Attention to the manner in which kings receive homage would result in kingly manners commanding homage. But only in king's houses are children taught to receive homage as kings. To be a noble courtier is the height of even the titled subject's ambition.

It was only as scion of the king's house that the little dauphin six years of age was seated on the French throne, with sceptre in hand and crown on head, that all the magnificent people of the court might pay him homage, that he might early learn to receive homage in kingly fashion.

At a certain point of attention to mathematics the origin of reckoning is touched, and man can perform any given calculation with numbers. At this height he spills over with mathematics. The very airs around him are instinct with his science. Omar Khayyam was one of the greatest mathematicians, and one of the greatest of observers of transactions by visioning; he wrote,

> "As when the tulip for its morning sup
> Of heavenly vintage from the ground looks up,
> Do thou devoutly do the same,
> Till heaven to earth invert you."

Kings of old sought such as could spill over or distil their knowledges into the atmospheres, to associate with the royal infants, that by psychometric encounter, or unconcious suggestion, the children of the royal house might learn easily, and thus easily outstrip all the children of the subjects of the realm.

Masters of art along any line affect the mystic brain films of those with whom they associate. How stimulating to the brain film is the aural spilling over of one who has been taught by Him of the Heavenly Heights, who saith, "Look unto Me," "I will instruct thee and teach thee," "I will . . . show thee great and mighty things."

What healing peace may circumjacently radiate from one who has acquainted himself with Him of the Heavenly Heights ever inviting, "Acquaint now thyself with (Me) , and be at peace."

What rest to the weary may radiate from one who has come into identification with the Heavenly Presence ever calling, "Come unto Me, and I will give you rest."

Is not Rest the greatest achieving power mentioned in the books of inspiration? "They rest from their labors; and their works do follow them." God is Rest. "Return unto Me," He saith. "Returning unto the Root is rest," wrote the Chinese sages.

Should a man be as rested as God he would do the works of God. The twelve labors of Hercules would be his daily accomplishment. Mythology veils these mystic lessons in story form. Whoever should rightly read mythology would find each god a noble lesson in Mystical Science. He would find the history of man and the cosmos lined out from start to finish.

Notice how Atalanta lost the race by saluting one little temptation en route. "Salute no man by the way," said Jesus. Notice how Thor is first among the gods to be destroyed in the twilight of the gods, the Armageddon of the last age. Thor of the ruthless hammer! Then the other gods also to be annihilated, step by step, till there be left only One God and His Name One, the New Name which no man knoweth save he that receiveth it. Is it not easy to pick out the god of each nation in this twilight of time?

Who is obeying the single Edict of the High and Lofty One inhabiting Eternity, "Look unto Me and be ye saved"—"And I will give you rest." "There remaineth therefore a rest to the people of God. For he that is entered into his rest, he also hath ceased from his own works, as God did from his. Let us labor therefore to enter into that rest."

The only labor for mankind is ofttime glancing up to the Lover ever with him, the Lord of hosts his name. "Zeus," said the Greeks, "Zeus is his name. He was, and is and ever shall be, the glorious of man and of the fruitful earth; therefore call upon Zeus."

Rest is relatedness to the Lord of effortless achievement. Its symbolic stone is amethyst. Rest that heals the sick and strengthens the weak is the happy arrival of one that hath chosen the God of Jacob for his strength, by choosing his name of achieving Rest—"Far above all principality, and power . . . and every name that is named."

The twelfth stone stands for happiness. The happy do not care what happens to them. They are the care free; the truly care-less. They have cast all their care on the Author of omnipotent energy. Men wot not what Zeus glory hides its beams in care-lessness, the outer significant husk of flawless, unweighted executiveness.

The symbolic stone of the happy care-free is the

amethyst, least precious hiding the most precious, the Benjamin stone among the jewels:: "The Lord shall cover him all the day long," said Jacob. It is the Joseph stone: Blessed is he, for his are the "chief things of the ancient mountains." It is the Gad stone, signifying one who executeth the justice of great Zeus, enlarging himself by recognizing his own centre as the glorious I AM: Highest God and Inmost God one I AM.

The happy are the hopeless. Over the gates of Dante's Inferno it was written, "Abandon hope." Over the gates of Paradiso it is also written, "They hope not." For do they hope in heaven? How can they hope when the hoped-for is come? Does the bridegroom hope for his bride when she is already his love, rest and home? Does the mother hope that she shall have a son, when it is already born and nursing at her breast? No, it is come. "Ye are come unto Mount Zion, . . . the heavenly Jerusalem, and unto the city of the living God, and to an innumerable company of angels; to the general assembly and church of the first born."

We are saved by hope," said Paul, "but hope that is seen is not hope. For what a man seeth, why doth he yet hope for?"

The symbolic stone of hope-lessness, advance signal of hidden love, rest, and home, is the amethyst. The amethyst serpent in the diadem of every Pharoah was symbolic of a secret royal and divine power. It often turned out to be rod of wretchedness for the weak as occult power exercised by kings of dark old Egypt. But the amethyst is now symbol only of the as-yet-unworded wisdom, true waiting rod of the energizing Christ Jesus Name, filled with deliverance for the soldiers of the Armageddon in the twilight of the gods. " (For) the name of the Lord is a strong tower; the righteous runneth into it, and is safe."

The amethyst stone is symbol of health-giving indifference. Indifference is a state hiding the potent sceptre of Almighty Zeus. God is the Great Indifferent leaving the world to lay hold of the beams and ethers of mystic healing, or let them go by. Our sun is the great indifferent, not caring whether it is ripening the grape or rotting the apple. Dorman was the great indifferent, unmindful of cries of pain, yet calming them by his marriage to the regulating Sunshine, glorious, peace-giving AMEN RA.

Indifference that achieves is the child of upward gazing. "And the government shall be upon his shoulder," wrote the prophet Isaiah. In the Cabala the intellectually mystical book of ancient Jewish lore and law, the child of the heavenly Dominant is born last in mankind; the crucifixion is first. If the honoring praise of our neighbors sets us up, the government is on the shoulders of praise and honor. If the hostility or hatred of our neighbors depresses us, the government is on the shoulders of hostility and hatred, two crucifying energies overpowering our powers. But if neither praise nor hostility obscures our sense of our I AM dominance, the government is on our shoulders, and by us our neighbors are set into their rightful relations with us, catching our principles of life and arriving side by side with us, neither condemning nor praising, but fellowshiping with us in glad hurrying toward the discovery of the New Science just at our gates, yet now hiding its splendid face from even the most advanced of mankind.

Virgil declared that the discoverers of new principles

wear white chaplets on their heads in the Elysian fields, to distinguish them from the common shades who never discovered a single one of the uncountable new laws that press against our mystic brain films.

There is a discovery for every mystic brain. There is a discovery for our fellowshiping brain; the crowning science, held in renitent hiding, till unwarring fellowship makes all mankind as one.

The amethyst signifies originality. It signifies the coming unity of mankind, that can by united drawing charm, raised to Nth power, pull hitherward the New Science that ushers in the millennium.

All who have discovered new principles have acknowledged that it was more as if they remembered the principles than as if they had never known them. Archimedes, Pythagoras, Euclid, remembered. ". . . if they had (only) been mindful," said Paul; (or remembered) "that country from whence they came out!"

Remembering is returning. And returning is repenting. Mary of Bethlehem remembered, returned, repented; the power of the Highest overshadowed her, and she brought forth Jesus, the only unhypnotized man that ever walked the planet. He was not hypnotized by the world belief in the drowning power of water, the incurableness of leprosy, blindness, deafness, death. He was not caught in the limitations of bread or water or gold. He set the Christian Religion in the earth, nearly a billion strong today, equal to victorious grappling with the powers and claims of all who oppose His love and peace doctrine, or forget His vicarious finished work.

We are all fated to remember, return, repent, and bring forth some mighty sign of our recognition of the overshadowing God. This is high light thrown on the doctrine of fate. Did not the ancient Romans declare that "None could breake ye chaine of forged destinie, that firme is tyde to Jove's eternal throne"?

Did not the Chinese of dim antiquity proclaim that each man is chained by a single golden thread to the flowery upper land of bliss? Did not they all declare that holding fast by our own upward drawing cord is clue to our own self-transcending, like to Jacob's holding fast the cord of his self-transcendence?

The sages of old India announced that memory is the thread that draws us upward to love, rest, and home. They remembered and remembered their sacred writings tome on tome. The early Christians insisted that by good works should man attain to heaven his longed-for home. But Jesus said love, rest, and home, should be ours by repentance in His name, beginning at Jerusalem, or each Self. As in the days of Augustus all roads led to Rome, even so do all the mystical lessons lead back again to repentance by Upward Glancing, and fidelity to the Revealing Name.

The Ophites and Nazarenes of old prophesied a wisdom religion to come to the world, having an ineffable Name for the secret of its triumphs. Tacitus and Seutonius declared that a Man from Judea should subjugate the nations. Who hath transcended Jesus of Nazareth, baptized with Christ the Victorious God, till death and hell and warfare can be glorified out of existence by the influence of His secret name ever breaking through his common name?

Who hath so transcended all laws of nature, rising from the dark tomb by identification with Christ the unburiable? Who being thus risen hath led his Apostolic worshippers as far as Bethany, and lifting up his

hands hath blessed them as a heavenly host on high, un-
noting of their estate as earth-born mankind?

Bethany means house of dates, or house of misery,
according as the vision is Godward or earth-bound.
Here again is the terrible alternative as led off by the
mystic sense of inward viewing.

Note how the Apostles being blessed as heavenly
host by the Risen Christ Jesus, returned to Jerusalem,
or began at their Jerusalem Self, shouting to the Most
High, "Blessing and honor and glory and power be unto
him that sitteth upon the throne . . ."

Note how they preached union with the Highest
and communion with the sons of light. They had been
taught that two are ever in the field, man and super-
man; one should be taken, the other left. Their address
was thus ever to the super-man, the Angel of man's
presence.

To the Angel of the amusement-bound, to them of
the Church of Ephesus let us write to Risen Christ. To
the Angel of the beauty-bound, to them of the Church
of Smyrna, let us write as to God, Origin of Beauty. To
the Angel of each church, let us write the great things
of The Highest.

To the Pergamos Angel let us write; to them of the
grandeurs of wealth, statecraft, learning; to the lovers
of art, architecture, music, war. To them let us write
as to One who transcends all grandeurs.

The Apostles wrote to the Pergamos type concerning
the true Healing God whom the Pergamites were worshipping in the temple as Æsculapius, the key god in-
visibly filling his outer temple, urging his own worship
with promises of health auras if mankind would only
marry unto him, the god of healing, radiating his sun-

shine through his temple's ivory walls, to envelop his devotees with resistless magistrum.

The Apostle wrote to the Angel of the Church of Pergamos, mentioning the truly victorious God, whom the Pergamites were adoring in the golden temple as Jupiter the Mighty. They wrote to the Angel in the Church of Pergamos, the Joyous God, whom the Pergamites were wooing in the shining temple as Bacchus the delectable. To the Angel of Pergamos, the city of splendid temples built of shittim wood and ivory, sandalwood and ebony, the Apostles of the Lord Christ of the one God wrote glowing letters.

Not to the people did they write, but to the Angel of the people. Herein is great light on the mystical ministry of the Messiah Christ. As he blessed his disciples as an angelic host, and thus made of them the transcendent among mankind, so all those who regard their neighbors as angelic host, above their seeming selves, imbue their neighbors with immuneness to disease, and hurting powers, and death, turning them to adoration of the High Redeemer.

The mystical books of the oracles could be found in Pergamos of old. To this day the subtle aromas of the books of the oracles are sending athwart receptive brain films their New Age wisdoms. For though the scrolls of high teaching be outwardly burned, their secret messages go ever stealing forth.

Though the Caliph Omar destroyed the scrolls of Pergamos and Alexandria, yet the writings on the scrolls are still whispering across our mystic brain films. Let us hearken for the high teachings with which the inner airs are charged. "The value of all the scrolls is in the Koran," shouted the Caliph Omar. And so it is.

Does not the Koran declare "Most Highest Allah," who teacheth what before mankind hath not known? Shall not the Master and the scholar who teacheth less than "Most Highest Allah" be cut off out of the tabernacles of Jacob? Is not our prophet Malachi inspired, when with foresight like to the makers of cosmic myths, sighting the end of the world's beloved gods, he sees that all of low viewings must be ended with their gods' great twilight?

When the Christian Theophilus, hurrying with his mob through Pergamos to burn the Sibylline prophetics, shouted that all books were as nothing in the light of Christian Scripture, was he not shouting better than we have credited to him? For what is the point of Christian Scripture now breaking in upon us, save, "Seek ye my face evermore"—"And every eye shall see, and every tongue confess." Did the Sibylline books go so straight to the saving instruction, though their secret doctrines read that Hecate, man's own I AM, shares equal honors with Zeus, the One I AM, because as there is but one I AM in the universe, I AM that I AM?

When Mark Anthony gave to Cleopatra 200,000 precious volumes from the library of Pergamos, was he not corralling heathen knowledge and mythological significance into one spot of old Africa, to silver the Christian gospel yet to be, with the wise secret potencies of hoary antiquity?

Shall not the Original Christian Gospel reach us in its own appointed time? Shall not the foretold Messenger arrive to whom even the Jews shall shout Hozannas? The Christ of the New Christian Gospel is not crucified, not buried, not risen. The Christ of the New Christian Gospel is the Able-to-save-from-falling, by the mighty

power of the Name within the name of that Wholly
Christ-imbued One, Bloom in the Garden of Time.

"Great is Diana of the Ephesians!" they shouted. But
Diana gave way to Mary. "Great is Mary!" they shouted.
But Mary gives way to the Holy Ghost. "Great is Jesus
the Crucified!" they shouted. But Jesus the Crucified
gives way to Uncrucified Free Christ.

Christ was never crucified. Christ is the uncrucifiable, unburied and unrisen—the Eternal and Changeless
Self-Existent High Redeemer—The Awakener of New
Powers and New Knowledges. Let the Name that hides
The Name of the Uncrucifiable Self-Existent ascend,
till its hidden manna nourishes our mystical bodies into
joyous prominence! Till the Holy Ghost that teaches
all things is our daily breath.

> Breath of heaven all truth revealing,
> Kindling in us life divine!

Only he of the mystical Christ Body Dominant can
be the writer of the New Book that is promised to ex-
pound the fundamentals of the New Dispensation and
explain the starry signs no other science seems to get
right—signs declaring that new heaven and new earth
now hurrying through the thundering Armageddon
gates.

"The stars in their courses fought against Sisera," because Sisera fought against the stars.

For the stars in their courses are a framework inevitable. They go according to Peace, and Order, and the
way of Kindness. Whatever opposes the purpose of the
order the stars do serve, must take the reactions of de-
fiance to their stately signallings.

For many years Sisera's kingdom had been working to overthrow the Jehovah Inexorable, as set forth by the Israelitish prophecies. His king Jabin had harassed Israel secretly and openly, and with stored munitions of iron chariots, nine hundred strong, had suddenly met outwardly unprepared Israel in battle array on the great plain of Esdraelon.

The framework of heaven gives to each star its path whereon not only to shine, but in its visible integrity to figure forth the Invisible integrity of a Finished Prototype.

Whatever people, or tribe, or country, or globe shall stand up for the Integrity that swings the stars, as according to loving kindness, and peace giving, and life-and-joy-defending, shall be, by so far as he thus stands up, in league with The Undefeatable.

How then could Deborah the prophetess, with upward vision, find any other prophecy for Barak of Israel, called the thunderbolt of the framework, than that Barak defence should stand in better with the stars than Jabin-Sisera onslaught?

"On earth peace, good will toward men," sang the angels, heralding One who never stepped out of league with the stars in their courses.

Then, if there is any nation or globe in the spaces, that is preparing to molest the peace of its neighbor, it is out of step with the heavenly legions, visible and invisible, and cannot win its iron-charioted way, even though all its munitions and plannings have risen to the Nth power of antagonism. Let not any such Sisera expect to triumph against the stars that wheel only Good Will.

Glance up, glance up often, O Man! So shalt thou

find thyself in unhinderable step with the Victorious Highest, Author of the Integrity that sends the stars on their sublime marches; Father of Him who gave His life to set mankind into rhythm with life everlasting; who gave His life to quicken our recognition of the Responsive Divinity overlooking the stars above the signal stars; the Responsive Divinity quiescent like a sleeping giant in the breast of every one of us; who gave His life to call our own attention to our own deeply held acquaintance with the Secrets of the Finished Universe, to the "light that lighteth every man that cometh into the world." Let us write to the Angel of our neighbor's presence: Ye are above the wheel of matter and the net-work of mind. Ye are light of the world—free, flawless, immortal.

INDEX

Above the pairs of opposites, 58, 129, 140, 152, 242, 270, 320
Above thinking, 58, 163
Abraham, inborn authority, 103
Absolute, 35, 55, 80, 113, 123. 134, 135, 207, 211, 239, 240, 245, 293
Acceptance—
  unceasing, 141
  world's, 198
Accomplishing executiveness, 240
Achievement—
  great, 95, 96
  planet of, 120
  princely, 51
  unital standard, 179
Achieving power, 2, 6
Acknowledge High Judgment, 298
Acknowledgment—
  awaiting owner's, 316
  finished work, 124
  of accomplishment, 113
  of Jesus Christ, 138
Actinic ray in sun, life-giving, 252
Adept, Jesus the greatest, 227
Adjustable, to finished estate, 312
Adversary, to pain, 61
Affairs, 157, 169, 273, 333
  healing of, 47
  newly right, 75
Alkahest—
  associate with, 50
  dissolving, 27, 32
  falling, 35
Alkahests—
  dissolve films, 13
  healing, 273
  soft, 47
  sprinkling, 44
All-accomplishing sense, 6
Angel—
  bright, 71
  fifth, 109
  first, 47
  of—
    deliverance, 207
    God, 163
    God's Presence, 110

His Presence, 163, 164, 177, 181, 185, 188, 199
  the miracle, 79
  the Presence, 242
  the Presence, or Judah Self, 245
  third seal, 50
  Victory, 81
  second, 40, 49
  Servant, 77
  spirits, 59
  third, 49
  told Ezekiel, 4
  wrestled with, 77
Angels—
  associating with, 51
  comrade of, 9
  comrade with, 49
  comraded by, 68
  ministerings of, 70
  of Apocalypse, perceptions, 76
  shall save, 57
Annointing, Christian formula, 247
Aphorisms, 11, 89
Appearance, 84, 108, 159, 177, 182, 196, 272, 288, 309
Appearances, 172, 217, 238, 258, 309
  caught in, 288
  Lord's ten, 159
Arcane way, 11
Assumption, taking for granted, 238
Assurance, 3, 40, 75, 79, 80, 115, 319
Atonement—
  day of, 61
  Man of, 149
Attention—
  awe-struck, 45
  exalted, saving effects, 15
  focussed, 53
  give, 31
  makes duplicate, 337
  Moses taught high, 164
  objective of, 53
  objective points to focus, 170
  objective worth our, 226
  persistent, 162, 205
  principle of, 83
  secret of success, 214
  Solomon's, 3

COSIMO is a specialty publisher of books and publications that inspire, inform and engage readers. Our mission is to offer unique books to niche audiences around the world.

COSIMO CLASSICS offers a collection of distinctive titles by the great authors and thinkers throughout the ages. At COSIMO CLASSICS timeless classics find a new life as affordable books, covering a variety of subjects including: *Biographies, Business, History, Mythology, Personal Development, Philosophy, Religion and Spirituality,* and much more!

COSIMO-on-DEMAND publishes books and publications for innovative authors, non-profit organizations and businesses. COSIMO-on-DEMAND specializes in bringing books back into print, publishing new books quickly and effectively, and making these publications available to readers around the world.

COSIMO REPORTS publishes public reports that affect your world: from global trends to the economy, and from health to geo-politics.

FOR MORE INFORMATION CONTACT US AT
INFO@COSIMOBOOKS.COM

If you are a book-lover interested in our current catalog of books.

If you are an author who wants to get published

If you represent an organization or business seeking to reach your members, donors or customers with your own books and publications

**COSIMO BOOKS ARE ALWAYS
AVAILABLE AT ONLINE BOOKSTORES**

VISIT COSIMOBOOKS.COM
BE INSPIRED, BE INFORMED